COLLINS POCKET REFERENCE

BIOLOGY

Michael Allaby
and
Michael Kent

D1579602

HarperCollins*Publishers*

HarperCollins Publishers
P.O. Box, Glasgow G4 0NB

First published 1996

Reprint 10 9 8 7 6 5 4 3 2 1 0

ISBN 0 00 470928 4

A catalogue record for this book is available
from the British Library

Printed and bound in Great Britain by
Caledonian International Book Manufacturing Ltd,
Glasgow, G64

Preface

This dictionary contains simple, straightforward explanations for more than 1500 terms that are widely used by biologists. In deciding which terms to include we have aimed primarily at the needs of A- and AS-level students taking any of the courses now available that have a strong biological component. The A- and AS-level biology syllabuses have undergone radical changes over the past few years. Modular schemes have widened the range of topics, but all syllabuses include a core of common topics. This dictionary covers all the main terms within the core. The nomenclature follows the recommendations made by the Institute of Biology and the Association for Science Education. Although the dictionary is intended primarily for A- and AS-level students it will also be of value to those working for the Scottish Higher Grade examination and the Certificate of Sixth Year Studies, to students of GNVQ Science and, indeed, to those who are not students at all in the formal sense, but who have a keen interest in biological topics.

We found it clearer to explain certain ideas in the form of one longer entry rather than in a number of shorter ones, so you will find some entries in the dictionary are much longer than others. Synonyms and terms defined within longer entries are indicated by italic typeface. Terms used within entries but defined under their own heading are cross referred in bold type.

At the end of *Pocket Reference Biology*, we have added appendices to supply further useful information in a more convenient and succinct form than would be possible in dictionary definitions. These describe the principal methods for displaying data graphically, the geological time scale, SI units of measurement, and the major groups of living organisms.

A dictionary is not a textbook, and should not be used as a substitute for one, but we hope this book will help you read textbooks more easily by providing simple, accessible explanations for the terms and concepts of biology that you may encounter.

Michael Allaby
Michael Kent

Wadebridge, Cornwall

Acknowledgments

Appendix 1, diagrams *bar chart, column graph, histogram, pie chart* from *Biological Nomenclature - recommendations on Terms, Units and Symbols*, The Institute of Biology

Appendix 3, diagrams from *Biology - A Functional Approach*, M B V Roberts, Nelson.

A band See *sarcomere*.

abdomen The posterior part of the body trunk of animals. In *vertebrates*, it contains the *kidneys, stomach, intestines* and reproductive organs. In mammals, it is separated from the *thorax* by the *diaphragm*. In most *arthropods*, it is segmented.

abiotic factor A non-living factor; abiotic factors form the non-biological part of the environment (e.g. inorganic nutrients, temperature, salinity, pH, availability of water, and light intensity).

abomasum See *rumen*.

abortion In animals and plants, the arrest of development of an organ. In mammals, the premature birth of a *foetus* (i.e. of one that is insufficiently developed to be able to live outside the womb). In humans, technically it is the expulsion of the foetus (spontaneously or induced) between the time of fertilization and the end of three months' gestation.

abscisic acid See *plant growth substance*.

absolute refractory period See *refractory period*.

absorption The uptake of energy or matter into a system (e.g. the uptake of light into chlorophyll, nutrients from the gut into the blood, and water and mineral salts from the soil into plant roots). Absorption of materials into cells may be active (requiring energy) or passive. See also *active transport, diffusion, endocytosis, facilitated diffusion,* and *osmosis*.

absorption spectrum A graph that shows the relative absorption by a substance of different wavelengths of radiation. *Chlorophyll* for example, has a characteristic absorption spectrum, absorbing mainly the red and blue wavelengths of the visible light spectrum, so it appears green.

accessory pigments See *photosynthetic pigments*.

accommodation 1. The decrease in responsiveness of a *neurone* when it is stimulated continuously. Accommodation (or fatigue) is due to exhaustion of *neurotransmitter* (see *synapse*). The neurone is able to respond again when it has had sufficient time without stimulation to synthesize more neurotransmitter. **2.** See *focusing*.

accuracy The closeness of a measurement to the true value of what is being measured. Using statistical analysis of a number of measurements, accuracy is often expressed as the deviation (+ or −) of the measurement from its true value. Accuracy is an important factor to consider when evaluating the results of an investigation.

acellular (of an organism or tissue) Not composed of separate cells, but which may have more than one *nucleus*. See also *unicellular organisms*.

acetic orcein A stain that is taken up by cell nuclei and commonly used to identify the different stages of *mitosis* in squashes obtained from *root tips*.

acetyl coenzyme A or **acetyl Co-A** An important two-carbon compound that acts as a link between *glycolysis* and the *Krebs cycle* in *aerobic respiration*. If oxygen is present, *pyruvate*, the end-product of glycolysis, passes into the *mitochondria* and is converted to acetyl Co-A. The two-carbon acetyl component of acetyl Co-A combines with the four-carbon compound oxaloacetic acid to form the six-carbon compound citric acid in the Krebs cycle. *Fatty acids* and *amino acids* must also be converted to acetyl Co-A before their energy can be released in the Krebs cycle and *electron transport system*.

acid A substance that gives up hydrogen ions (protons) readily. When dissolved in water, a *strong acid* (e.g. hydrochloric acid) is one that separates, or dissociates, into its component ions almost completely. Acids turn blue litmus paper red and have a *pH* of less than 7.

acid rain Precipitation with a *pH* lower than about 5.6, which is the usual value for rain into which natural constituents of the atmosphere (principally CO_2 and SO_4) have dissolved. The additional acidity is most commonly due to air *pollution*, the most important pollutants being carbon dioxide, sulphur dioxide, and nitrogen oxides, emitted as waste products from the burning of *fossils fuels*. Acid rain may harm freshwater *ecosystems*, the *soil* and trees. Acid mist, which clings to plant surfaces, and airborne acid particles deposited directly on leaves are more harmful to plants than acid rain.

acoelomate An animal (e.g. one belonging to the *Cnidaria* or *Platyhelminthes*) that does not possess a true *coelom*.

Acquired Immune Deficiency Syndrome See *AIDS*.

acrosome A fluid-filled sac at the front of the head of a *sperm* which contains powerful protein-digesting *enzymes*. These are released when the sperm makes contact with an egg cell and make an opening through the protective layers of the egg, thus enabling the sperm nucleus to gain entry into the *cytoplasm* and fertilize the egg nucleus.

actin A fibrous *protein* found in the contractile cells of all animals. It forms the microfilaments of the cytoskeleton of all eukaryotic cells (see *eukaryote*). Actin makes up the thin filaments in the I bands of *skeletal muscle*. See *sarcomere*.

action potential The transient change in potential difference across the membrane of a cell (usually a *neurone* or *muscle* cell) when the cell is stimulated. The resting potential, about −70 millivolts, is reversed, reaching about +30 millivolts, due to an influx of sodium ions. See also *nerve impulse*.

activation energy The amount of energy needed to start a chemical reaction. Even energy-releasing reactions (e.g. those of *respiration*) require activation energy to make molecules move faster, collide, and have a greater chance to interact. Activation energy can be applied externally as heat, but this could be dangerous for living organisms; they use *enzymes* which lower the activation energy for specific reactions and so accelerate them.

activator See *cofactor*.

active site A particular area of an *enzyme*, usually at its surface, which combines with a *substrate* to form a temporary enzyme—substrate complex during a reaction. The active site has a three-dimensional shape which fits that of the substrate. Anything which affects the shape of its active site thus affects the function of an enzyme. See also *denaturation*, *lock and key theory* and *induced fit model*.

active transport The movement of ions or molecules across a *cell membrane* against a *concentration gradient* (i.e. from a region of lower to a region of higher concentration). This requires the cell to expend energy, usually in the form of *ATP*. Active transport is stopped by metabolic poisons (e.g. cyanide) that prevent ATP production; it is also stopped in most tissues deprived of oxygen. Protein *carrier molecules* are probably involved in active transport.

adaptation 1. Any genetically determined, inherited characteristic that fits an organism to its environment, or the evolutionary process by which such a characteristic becomes common within a population of organisms (see *natural selection*).
2. A physiological or behavioural change in an organism exposed to certain environmental conditions, such that the organism improves its chances of survival in that environment. Such adaptations are temporary and not inherited.
3. A decrease in sensitivity of a sense organ or organism resulting from continuous or repetitive stimulation, so that increasingly intense stimuli are required to produce a particular response.

adaptive radiation An evolutionary process in which organisms sharing a common ancestor multiply and diverge to occupy different ecological *niches*. Adaptive radiation occurs when a species extends its range or colonizes a new environment that presents new opportunities and problems (e.g. the 14 species of finches on the Galápagos Islands are all thought to have evolved from one species that migrated there from the South American mainland; lack of competition from other birds enabled the ancestral finch to occupy a range of niches and diversify).

adenine See *purine*.

adenosine diphosphate (ADP) A single *nucleotide*, similar in composition to *adenosine triphosphate*, but with only two phosphate groups. ADP is formed when *adenosine monophosphate* combines with a second inorganic phosphate group; the acquisition of a third phosphate group converts ADP to ATP. The energy needed for this reaction is released during *respiration*. ADP and inorganic phosphate are produced when ATP is broken down by *hydrolysis* to supply *free energy* for the body's activities.

adenosine monophosphate (AMP) A *nucleotide* formed from adenine (base), ribose (pentose sugar), and a single phosphate group. AMP is sometimes formed when *adenosine triphosphate* (ATP) is hydrolysed (see *hydrolysis*) to release energy for metabolic activities. See also *cyclic-AMP*.

adenosine triphosphate or **ATP** A *nucleotide* consisting of adenosine (organic base), ribose (a five-carbon sugar), and three phosphate groups. It occurs in all living organisms, acting as the universal energy carrier. ATP is the only source of chemical energy which can be used directly in *metabolism*; all other sources of energy (e.g. *carbohydrates* and fats) must be converted to ATP before being used. ATP is synthesized from *adenosine diphosphate* (ADP) by *oxidative phosphorylation* during *respiration*. It is also synthesized during cyclic and non-cyclic *photophosphorylation* in the light stage of photosynthesis. In the presence of *adenosine triphosphatase*, a molecule of ATP is broken down by *hydrolysis* to ADP and inorganic phosphate; 30.6 kJ of energy are released which an organism can use almost instantaneously in energy-requiring reactions. ATP + water → ADP + P (and 30.6 kJ energy).

adenosine triphosphate

adenosine triphosphatase (ATPase) The *enzyme* which accelerates the breakdown of *adenosine triphosphate* (ATP) to *adenosine diphosphate* (ADP) and inorganic phosphate with the release of energy. It also accelerates the reverse reaction when ATP is synthesized from ADP, but this reaction consumes energy. ATPase tends to be concentrated in regions of a cell where it is in most demand (e.g. in the *cristae* of mitochondria and in the photosynthesizing membranes of chloroplasts).

ADH See *antidiuretic hormone*.

adhesion The force of attraction between molecules of different substances, holding them together (e.g. between *water* molecules and the walls of *xylem* vessels with which they are in contact). See also *transpiration*.

adipocyte See *adipose tissue*.

adipose tissue In animals, fairly loose *connective tissue* containing fat-storing cells (called adipocytes) that make up 90% of the tissue. Adipose tissue has a good blood supply. It tends to develop under the skin where it acts as a shock absorber, insulator, and energy store.

ADP See *adenosine diphosphate*.

adrenal gland A *hormone*-producing gland; in mammals, one adrenal gland is located anterior to each *kidney*. It is divided into an outer part (cortex), the inner part of which secretes steroids, and an inner part (medulla) which secretes *adrenaline*.

adrenaline or **epinephrine** A *hormone* secreted by the inner part of the *adrenal glands*. It has widespread effects on muscles, blood circulation, and carbohydrate metabolism, preparing the body for action, or what is commonly called the 'fight or flight response'. Adrenaline increases the heart rate, breathing rate and metabolic rate, and enables muscles to contract more forcefully.

adventitious root See *cutting*.

aerobic respiration A series of reactions requiring oxygen in which energy-rich compounds (e.g. *carbohydrates* and fats) are broken down to produce *adenosine triphosphate* (ATP). There are three stages in aerobic respiration: *glycolysis*, the *Krebs cycle*, and the *electron transport system*. During glycolysis, *glucose* is broken down to *pyruvate* (a three-carbon compound). This takes place with or without oxygen; if oxygen is available, the pyruvate is converted into *acetyl coenzyme A* (a two-carbon compound) which is fed into the Krebs cycle in the mitochondria. The main function of the Krebs cycle is to release from the *substrate* hydrogen atoms (an oxidation reaction) which are used in the electron transport system. During the Krebs cycle, carbon dioxide is also produced as a waste product and one ATP molecule is generated for each complete cycle of reactions. The hydrogen is fed into the electron transport system; this comprises a series of *carrier molecules* arranged in sequence on the inner membranes (*cristae*) of mitochondria. The hydrogen atoms (or electrons released from the hydrogen) are passed from one carrier to the next, resulting in a series of *redox reactions* that release energy used to synthesize ATP from ADP. Oxygen is the final acceptor of the hydrogen and electrons, so this process is known as oxidative phosphorylation. During the electron transport system 32—34 molecules of ATP are manufactured from each glucose molecule, giving a total of 36—38 molecules of ATP produced by aerobic respiration (according to how efficiently hydrogen removed during glycolysis is transported into the mitochondrion).

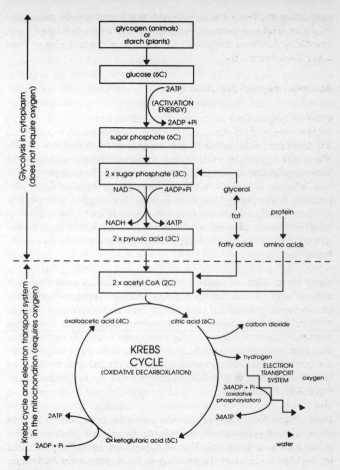

aerobic respiration (ATP production given per molecule of glucose.)

13

agglutination The clumping together of cells, usually resulting from an *antigen—antibody* reaction.

agriculture The systematic production of food and fibre (e.g. wool, flax) from the land by the raising of livestock and the growing of grass and field crops (e.g. cereals). The specialist growing of fruit, vegetables, and flowers is usually known as horticulture.

AIDS (Acquired Immune Deficiency Syndrome) An incurable human disease caused by a blood-borne *virus* (called human immunodeficiency virus, or HIV) that disrupts the functioning of the immune system. Death results from the inability of the body to fight infection. HIV virus is transmitted via body fluids, primarily through sexual relationships, and by infected blood or blood products used in transfusions or transferred when drug users share needles. It may also be transmitted by saliva.

algae (*sing.* **alga**) A wide range of plant-like organisms that inhabit aquatic or damp environments. Algae range from microscopic unicellular organisms to quite complex, large, multicellular structures, such as seaweeds. All algae have cells with *cellulose* cell walls and lack *vascular tissue*. The name has no strict taxonomic validity, the algae comprising a large number of phyla that are not necessarily closely related. In the five kingdom system of classification, these phyla are included in the *Protoctista*.

algal bloom A sudden, rapid increase in the population of *algae* in an aquatic *ecosystem*. In temperate regions, marine algal blooms occur naturally in the spring and autumn, the first coinciding with an increase in light intensity and the second with an upwelling of nutrients. Algal blooms can also be induced by nutrient enrichment (e.g. as a result of *eutrophication*).

alimentary canal or **gut** In animals, a tube in which digestion and absorption take place. In *Cnidaria* (e.g. *Hydra*), the gut consists of a simple sac with only one opening. The gut of *Platyhelminthes* (flatworms) also has only one opening, but it usually has many branches. In most other animal groups the gut is a tubular passage through which food enters at one end, the mouth, and is eliminated at the other, the anus. In mammals, the gut is differentiated into distinct regions which differ structurally and functionally. They are the *buccal cavity*, *oesophagus*, *stomach*, *small intestine* (duodenum and ileum), *caecum*, and *large intestine* which terminates at the *rectum*. Each region that digests food has a different pH, allowing specific digestive *enzymes* to work at their maximum efficiency.

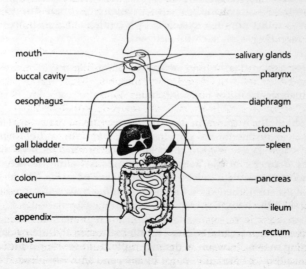

alimentary canal

alkali A base (e.g. sodium hydroxide) that dissolves in water. Solutions of an alkali have a *pH* of more than 7 and turn red litmus paper blue.

all-or-none law The law stating that a single nerve cell or muscle cell will respond to a stimulus (irrespective of its intensity) either completely or not at all. Thus there is no partial nerve impulse or partial contraction of a muscle cell.

allele See *gene*.

allelomorph See *gene*.

allergy The condition in which an individual is particularly sensitive (*hypersensitive*) to one or more substances that are harmless to most individuals. An allergy is often due to an overzealous immune system which treats such substances as though they were harmful *antigens*.

allopatric speciation See *speciation*.

allosteric inhibitor See *inhibition*.

alpha helix A common *secondary structure* of many *proteins* (especially the fibrous proteins, such as *keratin* and *collagen*) in which a polypeptide chain is twisted around its own axis. The diameter of the helix remains constant along its length, and each complete twist consists of 3.6 *amino acids*. The alpha helix is strengthened by *hydrogen bonding* between successive turns of the helix.

alternation of generations The alternation in the life cycle of an organism between a distinct diploid stage (see *diploidy*) that produces asexual spores by *meiosis*, and a haploid stage (see *haploidy*) that produces sexual gametes by *mitosis*. True alternation of generations occurs only in plants. The diploid spore-producing stage is called the *sporophyte*, and the haploid gamete-producing stage is called the *gametophyte*. Some animals (e.g. *Obelia* which belongs to the *Cnidaria*) have an asexual stage and a sexual stage, but both are diploid, so this is not true alternation of generations.

alveolus (*pl.* **alveoli**) See *lung*.

amino acid A building block of *protein*. All amino acids have an acidic carboxyl group (—COOH) and a basic amino group (—NH₂), making them *amphoteric*. The general structural formula for an amino acid is:

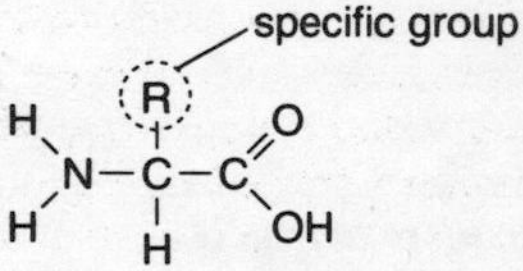

amino acid Structure of a generalized amino acid.

Where R represents a group that varies for each amino acid. For example, in the simplest amino acid, glycine, R = H. More than 170 amino acids occur in living cells, but only about 20 of these are used in human proteins. Some (non-essential) amino acids can be synthesized in the body, but others (essential amino acids) must be obtained from food.

Non-essential amino acids
   alanine (Ala)
   asparagine (Asn)
   aspartic acid (Asp)
   cysteine (Cys)
   glutamic acid (Glu)
   glutamine (Gln)

glycine (Gly)
hydroxyproline (Pro)
serine (Ser)

Essential amino acids
(obtained only from the diet)
arginine (Arg)
histidine (His) (essential in children only)
isoleucine (Ileu)
leucine (Leu)
lysine (Lys)
methionine (Met)
phenylalanine (Phe)
threonine (Thr)
tryptophan (Trp)
tyrosine (Tyr)
valine (Val)

ammonification See *nitrogen cycle*.

amniocentesis The removal of a small volume of the fluid surrounding a *foetus* in the *amnion* in order to obtain foetal cells for the prenatal diagnosis of genetic or biochemical defects (e.g. *Down's syndrome*). The fluid is withdrawn through a hollow needle inserted through the woman's abdominal wall and into the uterus. The procedure is usually carried out between the 16th and 20th week of pregnancy.

amnion A membrane that encloses the *embryo* of reptiles, birds and mammals. There is a space (*amniotic cavity*) between the membrane and embryo filled with a fluid (*amniotic fluid*) that protects the embryo from desiccation and cushions it from mechanical disturbances.

amniotic Pertaining to the *amnion*.

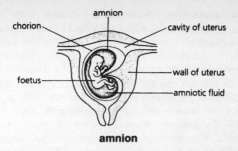

amnion

Amoeba A genus of *unicellular organisms* (kingdom Protoctista, phylum Rhizopoda) that form temporary *pseudopodia*, resulting in their constantly changing their body shape for movement (see *amoeboid movement*) and feeding. Amoebae are heterotrophic, engulfing prey (e.g. bacteria) with their pseudopodia by a process called *phagocytosis* ('cell eating'). The cytoplasm has no special sensory organelles, but it does possess a *contractile vacuole* for *osmoregulation*. The best known species, *Amoeba proteus*, reproduces asexually by simple *binary fission*. Different species of *Amoeba* occur in habitats ranging from the surface of teeth and the gut cavity to soil water, ponds and oceans.

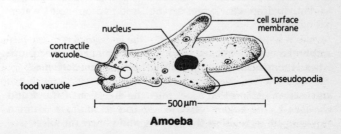

Amoeba

amoeboid movement Movement by means of the extension and retraction of *pseudopodia*, exhibited by *Amoeba* and some cells in multicellular organisms (e.g. mammalian *white blood cells*). During amoeboid movement, the fluid, inner layer of *cytoplasm* (called endoplasm) flows in the direction of movement into one or more pseudopodia. At the tip of the pseudopodium, the endoplasm is converted into a more viscous form of cytoplasm called plasmagel. This process is reversed at the trailing end as the outer cytoplasm (ectoplasm) is converted into endoplasm. It is thought that the contractile proteins *actin* and *myosin* are involved in producing the motive force required for amoeboid movement.

AMP See *adenosine monophosphate*.

amphetamines A group of drugs, similar to *adrenaline*, that act as powerful stimulants on the *central nervous system*.

Amphibia or **amphibians** A class of soft-skinned, chordate animals (see *Chordata*), including frogs, toads and newts. They usually have an aquatic, larval stage with gills (when they are known as tadpoles), and a terrestrial, adult stage with lungs.

amphibians See *Amphibia*.

amphipathic molecule A molecule (e.g. a *phospholipid*) that has both hydrophilic ('water-loving') and hydrophobic ('water-hating') ends.

amphoteric (of chemicals) having both acidic and basic properties. *Amino acids* are amphoteric. See also *zwitterion*.

amylase An *enzyme* that accelerates the digestion of *starch* and *glycogen* to the sugars *glucose*, dextrin or *maltose*. Human amylases are secreted in the salivary and pancreatic juices (see *pancreas*). Plant amylases mobilize starch food stores in germinating seeds and underground storage organs such as *rhizomes* and *tubers*.

amylopectin A branched-chain polysaccharide found in natural *starches*.

amylose A component of natural *starch*, consisting of a long, unbranched chain of *glucose* units.

anabolic reaction An energy-requiring reaction in a living organism that results in the synthesis of complex molecules from simple ones. Examples include photosynthesis, *protein* synthesis, *nucleic acid* synthesis and polysaccharide synthesis. The sum of all the anabolic reactions in an organism is called anabolism. Compare *catabolic reaction*.

anaemia A deficiency in the number of *red blood cells* or *haemoglobin* in the blood. This reduces the ability of the blood to carry oxygen and is characterized by shortness of breath and headaches. One form (pernicious anaemia) is caused by lack of *vitamin B12*, other forms by lack of iron in the diet, an inability to absorb iron, or excessive bleeding. See also *sickle cell anaemia*.

anaerobe See *anaerobic respiration*.

anaerobic respiration *Respiration* that occurs in the *cytoplasm* in the absence of oxygen. It may include the breakdown of energy-rich compounds (e.g. creatine phosphate) to make *ATP*, but most anaerobic respiration involves *glycolysis*, resulting in the breakdown of *glucose* to pyruvic acid (= *pyruvate*). Glycolysis provides a quick source of ATP, but each glucose molecule has a net production of only two ATP molecules. In plants and fungi, the pyruvic acid is converted to ethanol and carbon dioxide (a process called alcoholic fermentation). In animals, it is converted to *lactic acid*. Organisms that respire anaerobically are called *anaerobes*.

analagous structure A structure or organ in one species that performs a similar function to one in an unrelated species, but differs in its detailed structure and origin. Analogous structures appear superficially similar (e.g. the wings of insects and the wings of birds).

anaphase The stage in nuclear division in which either *chromosomes* (in the first division of *meiosis*) or sister *chromatids* (in *mitosis* and the second division of meiosis) move to opposite poles of the cell by the action of the *spindle apparatus*.

anatomy The study of the form of organisms and of those structures that can be revealed by dissection.

androecium The collection of stamens within a *flower*; together they constitute the male reproductive structures.

aneuploidy See *mutation*.

Angiospermae See *Angiospermatophyta*.

Angiospermatophyta, **angiosperms** or **flowering plants** A phylum (or division) of vascular plants (tracheophytes) that bear true flowers which produce seeds enclosed in a fruit formed from the ovary. There are two classes: Monocotyledonae (monocots) and Dicotyledonae (dicots). In some classifications the flowering plants rank as a subphylum (subdivision), Angiospermae. See *dicotyledon* and *monocotyledon*.

angiosperms See *Angiospermatophyta*.

animal husbandry The commercial breeding and raising of farm animals, nowadays usually employing scientific methods.

Animalia or **animals** A kingdom consisting of multicellular organisms with *nervous coordination* that do not photosynthesize and that exhibit *heterotrophic nutrition*. See also *five-kingdom classification*.

animals See *Animalia*.

animal starch See *glycogen*.

Annelida or **annelids** A phylum of worm-like animals in which the body is divided externally and internally into well-defined segments. There are three main classes. Polychaetae (bristle worms, e.g. *Nereis*, ragworms) are marine. Typically they have a distinct head, numerous *chaetae* (retractible bristles), and paired, fleshy, paddle-like flaps of tissue (called *parapodia*) on each segment, used for locomotion. Oligochaetae (oligochaetes or earthworms, e.g. *Lumbricus*) occur mainly in soil and fresh water. They have no distinct head, few chaetae per segment, and no parapodia. Hirudinea (leeches, e.g. *Hirudo* the medicinal leech) are usually ectoparasites (see *parasite*) or predators. They have no distinct head, no chaetae and no parapodia.

annelids See *Annelida*.

annual An organism that completes its life cycle in one year. The term usually refers to plants (e.g. the sunflower) that germinate, flower, produce seeds and die in the same year.

antagonistic muscles A pair of muscles that have opposite actions; the contraction of one prevents the contraction of the other and while one of the pair is contracting the other muscle is relaxed. Because muscles can only pull, a pair of muscles is required to extend, bend, or flex a joint (e.g. the biceps and triceps muscles in the arm act as an antagonistic pair: contraction of the biceps flexes the arm and contraction of the triceps extends it).

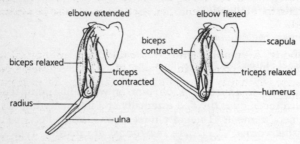

antagonistic muscles

antenna A long, whip-like, jointed appendage on the head of many *arthropods*. Antennae usually have a sensory function and contain receptors for touch and taste.

anterior Situated at the front (head end) of a person, organ, or part of an organism.

anther See *flower*.

antheridium The male reproductive organ in algae, bryophytes (liverworts and mosses), ferns, and fungi. It produces small, motile, male *gametes* called *antherozoids*.

antherozoid A motile *gamete* produced within an *antheridium* of algae, liverworts, mosses, and ferns. Antherozoids are *haploid* and move by means of flagella (see *flagellum*). The number of flagella and shape of the antherozoid vary from species to species.

antibiotic A substance that selectively kills or prevents the growth of microorganisms and is usually produced by another microorganism; for example, the fungal mould *Penicillium* secretes the antibiotic penicillin. Antibiotics can also be synthesized.

antibody A complex *protein* produced by an animal in response to the presence in its body of a foreign substance (an antigen) which may be harmful. The body can make an almost unlimited variety of different antibodies; each is highly specific and recognizes only a particular foreign substance or substances very closely related to it. The antibody binds to the foreign substance in order to kill it or render it harmless. In mammals, antibodies are produced by *plasma* cells derived from special *lymphocytes* (called beta-lymphocytes or B-cells), small cells originating in the bone marrow and found in large numbers in the *lymphatic system* (especially lymph nodes), spleen, and blood plasma.

anticoagulant A substance that prevents or slows down the formation of a blood clot.

anticodon See *RNA* and *artificial selection*.

antidiuretic hormone (ADH) A mammalian *hormone*, secreted by the posterior part of the *pituitary* gland, that plays an important role in *osmoregulation* by increasing water reabsorption from the *kidney*. ADH is thought to increase the permeability of the walls of the distal convoluted tubules (see *nephron*) and collecting ducts in the kidney. ADH deficiency results in a disorder called diabetes insipidus in which large volumes of urine are eliminated.

antigen A substance (often a *protein*) which triggers the production of an *antibody*.

antiseptic Any substance that kills or inhibits the growth of harmful microorganisms. Hydrogen peroxide and ethanol (alcohol) are common components of antiseptics used to treat minor wounds.

anus The posterior opening of the alimentary canal through which faeces are eliminated at intervals. In mammals, the anus can be closed by a ring of muscles (the anal *sphincter*).

anvil See *incus*.

aorta The large *artery* in mammals, that leaves the heart from the left ventricle. The aorta branches to form smaller arteries that carry oxygenated blood to all parts of the body.

apical dominance A condition in plants, in which the buds at the tip of a stem grow but lateral ones do not, encouraging the plant to grow tall rather than bushy. See *plant growth substances*.

apical meristem See *meristem*.

apodeme See *exoskeleton*.

appendage A projection from the body of an animal (e.g. antennae, legs and parapodia).

appendix A worm-shaped extension to the *caecum* of the mammalian gut. In herbivores it contains symbiotic bacteria that play a vital role in digesting plant *cellulose*. In humans, the appendix seems to have no function.

aquatic Relating to watery environments or organisms which live mainly in water.

aqueous humour See *eye*.

arachnids See *Arthropoda*.

archegonium A multicellular, flask-shaped structure, with a narrow neck and a swollen base, which contains the female *gamete* of bryophytes, ferns and many conifers.

areolar tissue See *connective tissue*.

arithmetic mean See *mean*.

arteriole See *artery*.

artery A tubular vessel that carries blood from the heart to another part of the body. Apart from the pulmonary artery, which carries deoxygenated blood to the lungs, and umbilical arteries, which carry deoxygenated blood to the *placenta*, all arteries convey oxygenated blood. They have thick, elasticated, muscular walls and carry blood under high pressure. As they approach a target organ, arteries branch to form smaller arterioles, some of which have small *sphincter* muscles in their walls that enable them to control the blood flow through the organ.

Arthropoda or **arthropods** Animals that have a hard *exoskeleton*, a bilaterally symmetrical body (see *bilateral symmetry*) divided internally and externally into clearly visible segments, and jointed limbs. The group includes the crustaceans, characterized by having two pairs of antennae and a head that is not clearly defined, the arachnids (spiders, mites and scorpions) which have four pairs of legs and no true jaws, and *insects*.

arthropods See *Arthropoda*.

articulation See *joint*.

artificial insemination The introduction of *semen* into the reproductive tract of a female mammal by artificial means (usually from a syringe introduced at the mouth of the *uterus*) so that only semen from a selected donor can fertilize an ovum. Artificial insemination is used as a part of *artificial selection* in domesticated animals, and in humans in some cases of infertility or impotence.

artificial selection The selection by humans of particular individual organisms for breeding. The organisms which best show desired characteristics are allowed to interbreed and organisms lacking desired characteristics are prevented from breeding (e.g. by killing, segregation or sterilization). This special form of *directional selection* gradually alters the average *genotype* of a group of organisms, forming new breeds and varieties. The wide variety of domesticated animals and plants alive today is largely a result of artificial selection which has been taking place since prehistoric times.

ascorbic acid See *vitamin C*.

asepsis The state of being free from disease-causing microorganisms. Asepsis is important in hospital operating theatres and when performing certain microbiological investigations. Techniques used to create asepsis (aseptic techniques) include filtering air, sterilization of all clothing and instruments, and the use of protective clothing (e.g. face masks and gloves).

aseptic technique See *asepsis*.

asexual reproduction A process by which a single parent gives rise to offspring by mitotic division (see *mitosis*) without the formation and fusion of *gametes*. Since only one individual is involved, asexual reproduction has the advantage that no partner needs to be found. It also enables a population to grow very quickly if the environment is suitable. Unless *mutations* occur, however, the daughter cells are genetically identical to the parent cell. Although this is advantageous in a stable, unchanging environment, lack of variation may make it difficult for a species to adjust to a change in the environment. Types of asexual reproduction include binary fission, *budding* and *vegetative reproduction*.

assimilation 1. The uptake of elements and simple inorganic compounds in autotrophs (see *autotrophic nutrition*), and their incorporation into complex organic compounds. Plants, for example, assimilate carbon dioxide and water into glucose). **2.** The uptake into cells, in heterotrophs (see *heterotrophic nutrition*), of the products of digestion, and their incorporation into cellular structures and products.

atom The smallest part of an element that can take part in a chemical reaction.

ATP See *adenosine triphosphate*.

atrium See *heart*.

average See *mean*, *median* and *mode*.

auditory nerve In mammals, a sensory nerve carrying information about the pitch and intensity of sound from the *cochlea* to the cerebral cortex of the brain, and information about posture and movement from the *semicircular canals* to the *cerebellum*.

autecology See *ecology*.

autoclave A device used for the *sterilization* of instruments and other materials which exposes them to steam at high pressure.

autonomic nervous system See *nervous system*.

autoradiograph A photographic film that records the image of a radioactively-labelled specimen (e.g. large molecules, cell components or body organs) when the specimen is rested on the film in the dark.

autosomal inheritance The inheritance of characteristics determined by genes which are carried on autosomes (i.e. any *homologous chromosomes* which are not sex chromosomes and which are identical in appearance). Compare *heterosomal inheritance*.

autosome See *autosomal inheritance*.

autotroph See *autotrophic nutrition*.

autotrophic nutrition A form of nutrition in which organisms (called *autotrophs*) use inorganic materials to make complex organic products. The chief sources of carbon and nitrogen are carbon dioxide and nitrates respectively. The energy for the synthesis of organic products may come from sunlight in photoautotrophs or from chemical reactions in chemoautotrophs.

auxin See *plant growth substance*.

Aves or **birds** A class of animals (phylum Chordata) that have wings, skin covered in feathers, and lungs. They produce eggs with shells and are endothermic (see *endotherm*).

axil See *axillary bud*.

axillary bud In plants, a bud that develops from the axil (i.e. the angle between the upper side of a leaf and the stem on which the leaf grows).

axon A cytoplasmic extension of a nerve cell that carries a nerve impulse away from a cell body. In large mammals, an axon of a motor *neurone* may be several metres long.

bacillus (*pl.* **bacilli**) A rod-shaped *bacterium*. Bacilli may occur singly or in chains. Those that occur singly include *Escherichia coli*, which inhabits the human gut, and motile nitrifying bacteria (e.g. *Nitrobacter* and *Nitrosomonas*), which inhabit the soil. Another soil bacillus, nitrogen-fixing *Azotobacter*, occurs in chains.

backbone See *vertebral column*.

back cross A *cross* (mating) between a parent and one of its own offspring. See also *test cross*.

bacteriochlorophyll See *chlorophyll*.

bacteriophage A *virus* that attacks *bacteria*. Each bacterio-phage is specific for only one type of bacterium.

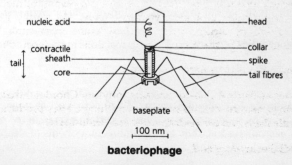

bacteriophage

bacterium (*pl.* **bacteria**) The smallest organism to have a cellular structure. Bacteria range in size from about 0.1 to 10 μm and belong to the kingdom *Prokaryotae*. A bacterial cell has a rigid *cell wall* made of *murein* (a molecule consisting of parallel chains of *polysaccharide* cross linked to form short peptide chains), sometimes covered with gummy secretions (see *cell capsule*). Bacteria may be *autotrophic* or *heterotrophic*. Bacteria which photosynthesize were formerly called 'blue-green algae'. They are now regarded as a special group of bacteria, called *cyanobacteria* (or blue-green bacteria). Non-photosynthetic bacteria are classified traditionally according to several criteria, including size, shape, motility and whether they occur singly or in chains (see also *bacilli*, *cocci*, *spirilli*, and *Vibrio*). Bacteria may be aerobic or anaerobic (see *aerobic respiration* and *anaerobic respiration*). They lack *mitochondria* and other organelles bounded by membranes and they contain no large vacuoles. Bacterial DNA is in the form of a large circular molecule and smaller *plasmids*; there is no true *nucleus*. All bacteria can produce asexually by *binary fission* but some types may also reproduce sexually. Bacteria are present in all parts of the *biosphere*. They may be free-living, parasitic or symbiotic. Bacteria are very important to humans: in addition to playing a central role in *nutrient recycling*, some cause disease in animals and plants, and others can be utilized in economically important processes, including *genetic engineering*.

balanced diet See *diet*.

bar chart See appendix 1.

barbiturate Any one of a group of drugs that act as powerful sedatives by reducing the activity of certain nerves in the *central nervous system*; they are common components of sleeping tablets.

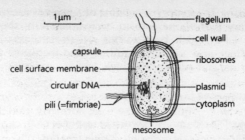

bacterium Generalized structure.

bark The protective, waterproof layer of woody stems and roots, consisting of three layers external to the *xylem*:
(a) an inner layer that includes living phloem cells;
(b) a middle layer of cork cambium (cells which produce cork cells by mitosis);
(c) an outer layer consisting mainly of dead cork cells impervious to air and water. Small pores (see *lenticels*) penetrate the bark, allowing gas exchange to take place between the atmosphere and the internal tissues.

basal metabolic rate (BMR) The lowest rate of *metabolism* (or energy utilization) at which a non-feeding, resting organism can remain alive in an environment at the same temperature as its body. BMR is usually expressed in terms of energy per unit surface area per unit time and varies with the size, sex, and age of the organism.

base A substance that reacts with an acid to form a *salt* and water only. Bases are sometimes called proton acceptors because they readily combine with hydrogen ions. Most bases are insoluble in water. Organic bases include the *purines* and *pyrimidines* which are components of *nucleotides*. See also *alkali*.

base pairing The *hydrogen bonding* of a *purine* base to a complementary *pyrimidine* base in *nucleic acids*. In the double helix of DNA, such base pairing occurs between cytosine (pyrimidine) and guanine (purine), and between thymine (pyrimidine) and adenine (purine).

batch processing See *fermentation*.

behaviour The externally observable response of an organism to an internal or external stimulus. See also *instinct* and *learning*.

belt transect See *transect*.

Benedict's test A test for *reducing sugars* (e.g. *glucose*). Typically, a 2 cm^3 solution to be tested is added to an equal volume of Benedict's solution; the solutions are mixed and heated gently. The formation of a coloured precipitate indicates the presence of reducing sugar; the colour of the precipitate changes from yellow-green to red as the concentration of the sugar increases. Benedict's solution is alkaline, transparent blue, and contains copper (II) sulphate ($CuSO_4$), sodium citrate, and sodium carbonate. Reducing sugars contain either an aldehyde group (–CHO) or a ketone (>C=O) group which can reduce the copper II ions (Cu^{2+}) to copper (I) ions (Cu^+); the copper oxide so formed is a brick-red colour. Benedict's solution can also be used to test for non-reducing sugars if these are first converted to reducing sugars by boiling them in hydrochloric acid. After boiling, the test solution must be cooled and neutralized (e.g. with sodium hydrogen carbonate).

beriberi A human disease caused by *vitamin B1* (thiamine) deficiency. Beriberi is characterized by decreased appetite, muscle wasting, nervous disorders and heart-muscle enlargement which can be fatal. (Sinhalese *beri* means 'weakness'.)

beta-pleated sheet A structural arrangement in *proteins* in which parallel polypeptide chains are cross-linked by *hydrogen bonds*. This forms a very strong structure. An example is silk.

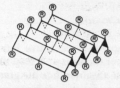

beta-pleated sheet Three parallel polypeptide chains with the imaginary pleated sheet (shaded) between them.

bicuspid valve See *heart*.

bilateral symmetry The arrangement of the body components of an organism such that a cut along a single imaginary line (usually running down the midline, through the mouth and digestive cavity), divides the body into two halves which are approximate mirror images of each other. Bilateral symmetry in animals is an adaptation to active movement in which opposing muscles lie either side of the body (see *antagonistic muscles*). It is also associated with *cephalization*.

bilayer See *fluid-mosaic model*.

bile A thick, brown-green, alkaline secretion from the *liver* that contains inorganic salts, bile pigments (derived from the breakdown of *haemoglobin*), and *cholesterol*. Bile helps emulsify fats (see *emulsification*), increases the activity of certain **enzymes**, increases the absorption of certain *vitamins*, and helps neutralize stomach acids.

binary fission A form of asexual reproduction in which a cell divides to produce two genetically identical daughter cells. First the nucleus divides by *mitosis* and then the *cytoplasm* splits into two by *cytokinesis*.

binocular vision See *vision*.

binomial nomenclature A system of naming organisms first devised by Carolus Linnaeus (1707—78). Each organism is given two names. The first name is the generic name and refers to the genus to which several distinct but similar species of organism may belong. It is given an initial capital letter. The second name is the trivial name (= specific name). It is always used in conjunction with the generic name to identify a particular species of a genus, and both names are italicized or underlined to distinguish them from common names. For example, the scientific name for humans is *Homo sapiens*. The generic name is sometimes used on its own when referring to characteristics shared by the whole genus. The trivial name is never used by itself. The allocation of scientific names (generic name plus trivial name) of organisms is governed by strict International Codes of (Botanical and Zoological) Nomenclature. Each two-part name is unique to a single species of organism and is used in all languages throughout the world.

bioaccumulation See *DDT*.

biochemical evolution The processes by which *biological molecules* formed from non-living matter and acquired the ability to combine and carry out the functions of living organisms (e.g. *metabolism* and replication).

biochemical oxygen demand, biological oxygen demand or **BOD** A measure of organic pollution in water; it is the amount of oxygen lost from a known volume of water kept in darkness at 20 °C for five days, usually measured in milligrams per cubic decimetre. The oxygen loss is assumed to be due to micro-organisms decomposing organic material. A low BOD therefore indicates low pollution levels.

biochemistry The chemistry of living things; the study of that chemistry, especially of the structure and function of the chemical components of organisms (mainly *carbohydrates*, *lipids*, *proteins* and *nucleic acids*).

biocontrol See *pest*.

biodegradation The breaking down of inorganic and organic substances by living organisms (usually involving *bacteria* or *fungi*). Biodegradation of biological substances is called *saprobiontic nutrition*.

biogas Combustible gas, usually containing methane, produced by the fermentation of organic matter. It can be produced in fermenters to provide fuels, but is also produced under compacted waste in some rubbish dumps, where it can form a dangerous, explosive mixture.

biological clock An internal, physiological, time-keeping system that underlies rhythmic patterns of *metabolism* and *behaviour* such as *circadian rhythms* and *photoperiodism*).

biological control See *pest*.

biological magnification See *DDT*.

biological molecules The molecules of which animals and plants are made.

biological oxygen demand See *biochemical oxygen demand*.

biology The study of life and of all the structures and processes associated with living organisms.

biomass, standing crop or **standing stock** The total amount of living substance in the organisms being studied, usually expressed in units of dry mass per unit area of the environment in which the organisms live, although it can be expressed in units of volume, mass, or energy and may refer to the whole or part of an organism.

biome A part of the *biosphere* that consists of a large area of land with similar environmental conditions (especially climate) and characteristic plants. Examples are desert and tropical rain forest.

bioreactor See *fermentation*.

biosphere That part of the Earth's surface, atmosphere, and waters in which organisms live.

biotechnology The use of organisms, their parts, or their products for practical human benefit. In its widest sense, biotechnology includes a wide range of practices, some of which are ancient (e.g. bread-, wine- and *cheese-making*, and the selective breeding of domestic animals and plants), some of which are only decades old (e.g. penicillin production), and some of which are still at an early stage of development (e.g. genetic engineering, and the production of *monoclonal antibodies*). Recently, biotechnology has tended to acquire a more restricted meaning, referring only to the production of genetically modified cells and microorganisms (especially bacteria).

biotic factor An environmental factor that results from the activities of living organisms. Biotic factors include the relationships between organisms (e.g. competition for food, predator—prey relationships, parasite—host relationships, etc.). Compare *abiotic factor*.

biotin A water-soluble *vitamin* belonging to the vitamin B complex, made by intestinal bacteria. It acts as a coenzyme (see *cofactor*) in the *metabolism* of *carbohydrates*, fats and *proteins*. Deficiency causes dermatitis and intestinal problems. Rich dietary sources include liver and yeast.

birds See *Aves*.

birth The act of bringing forth young from a female animal. The term is usually applied to *viviparous* organisms (i.e. those, such as mammals, that produce active young).

Biuret test A test for *proteins*. Sodium hydroxide (or potassium hydroxide) is added to an equal volume of test solution. A few drops of 1% copper sulphate is then added and mixed. Copper ions form a complex with the nitrogen atoms in a peptide chain. A purple coloration indicates the presence of *peptide bonds*; if no peptide bonds are present the test solution remains blue.

bivalent One of the pairs of *homologous chromosomes* which come together during *prophase* of the first division of *meiosis*. A bivalent consists of four sister *chromatids*.

Bivalvia The former name for Pelecypoda. See *Mollusca*.

bladder 1. An extensible, muscular sac into which the *ureter* continuously passes urine from the *kidney* and where urine is stored. When full, the urine is eliminated through the *urethra*

to the outside world. Emptying of the bladder is controlled by a sphincter muscle between the bladder and urethra.
2. An air-filled sac within the fronds of some seaweeds (e.g. the bladder-wrack, *Fucus vesiculosus*) that provides buoyancy.

blind spot The area of the *retina* from which the optic nerve leaves the *eye* and on which no image is formed, because it contains no rods and cones and is thus insensitive to light.

blood A watery body tissue in animals, that acts as a transport medium. Mammalian blood is contained within blood vessels and is circulated around the body by contractions of the *heart*. The blood is composed of cells or cell fragments suspended in a clear, straw-coloured fluid matrix called plasma. The plasma is about 90% water and 10% other inorganic and organic compounds; it transports carbon dioxide (mainly as hydrogen carbonate ions), heat, *hormones*, nutrients, and nitrogenous wastes (mainly *urea*). The cellular components comprise about 45% of the volume of blood and there are three main types: *red blood cells* (*erythrocytes*); *white blood cells* (*leucocytes*); and *platelets* (cell fragments). Red blood cells contain *haemoglobin*, which carries oxygen from the lungs to respiring tissues. White blood cells form an important part of the body's defence system. Platelets are involved in *blood clotting*.

blood clotting The process by which *blood* coagulates to form a solid plug (clot) made of blood cells trapped in a fibrous network. When a wound bleeds, the clot prevents further loss of blood, reduces the risk of pathogens entering the body, and provides a scaffold for the repair of damaged tissue. Blood clotting involves a complex series of biochemical reactions. When a wound occurs, blood *platelets* stick to exposed *collagen* fibres. The platelets release thromboplastin (Factor X, a blood-clotting factor) which works with calcium ions to convert *prothrombin* (an inactive plasma protein) to *thrombin* (an active plasma protein). Thrombin acts as an *enzyme*, catalysing the conversion of soluble *fibrin* into insoluble *fibrinogen*. Fibrinogen forms a network of fibres that traps blood cells and debris to form the clot.

blood glucose *Glucose* dissolved in blood plasma. In humans, glucose forms about 80% of the total blood sugar and is the only substrate (under normal circumstances) used by the brain as a source of energy. Blood glucose levels are kept relatively constant, in the range 3.4—5.6 millimoles per litre (= 60—100 mg per decilitre), by a complex homeostatic device (see *homeostasis*) that uses *insulin* and *glucagon*. Too much blood glucose (hyperglycaemia) or too little (hypoglycaemia) can be harmful. See also *diabetes mellitus*.

blood plasma See *blood*.

blood pressure The force exerted by blood on the walls of a blood vessel through which it is passing. In humans, blood pressure is greatest during contraction of the ventricles of the heart (systolic pressure), which forces blood into the arteries, and is lowest when the heart is relaxed (diastolic pressure) and filling up with blood. Blood pressure is measured in millimetres of mercury and is usually expressed as two numbers (e.g. 120/80, the normal blood pressure for a healthy adult person). The first figure refers to the systolic pressure, the second to diastolic pressure. Abnormally high blood pressure (*hypertension*) may result from partial blockage of the arteries or it may occur without a known cause.

blood vessel A tubular structure through which the *blood* of an animal flows. See *artery*, *capillary* and *vein*.

blow pooter See *pooter*.

blue-green algae See *bacterium*.

BMR See *basal metabolic rate*.

BOD See *biochemical oxygen demand*.

Bohr effect or **Bohr shift** A shift to the right of the *oxygen dissociation curve* of haemoglobin due to an increase in the concentration of carbon dioxide and a lowering of pH in the blood. At most oxygen *partial pressures* in the body, this results in a lowering of the affinity of haemoglobin for oxygen and a release of oxygen from oxyhaemoglobin. The Bohr effect occurs in blood supplying respiring tissues which produce high levels of carbon dioxide, and it results in oxygen being passed more easily into the tissues.

Bohr shift See *Bohr effect*.

bolus A rounded mass of chewed food softened by saliva and made suitable for swallowing.

bone marrow Soft tissue within the internal spaces of bones (e.g. the central cavity of long bones). In young animals, bone marrow is red and produces blood cells. In mature animals, red bone marrow is confined to the spaces within spongy bone. The large cavities of long bones contain mainly fatty tissue called yellow bone marrow.

bony fish See *Osteichthyes*.

boom or bust See *J-shaped growth curve*.

Bowman's capsule See *nephron*.

brain A concentration of nervous tissue at the *anterior* end of an animal which acts as the coordinating centre of the nervous system. In vertebrates, it consists of a highly organized mass of

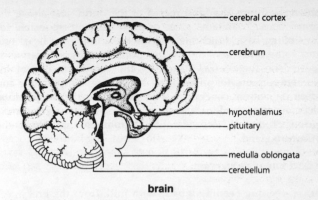

cerebral cortex

cerebrum

hypothalamus
pituitary

medulla oblongata
cerebellum

brain

nerve cells formed as an enlargement of the spinal cord and encased within the cranium of the skull. The brain analyses and integrates incoming sensory information and generates messages that control the activity of muscles and some other *effectors*. The brain also acts as a memory store. The main parts of an adult vertebrate brain are the *cerebrum*, *cerebellum*, *medulla oblongata* and *hypothalamus*. The cerebrum is the largest part. It consists of two hemispheres of nervous tissue (cerebral hemispheres), the surface of which is highly folded grey matter containing billions of nerve cells. In mammals, the cerebrum is the main integration centre of the brain, and contains sensory and motor areas for different parts of the body. It is also the seat of intelligence and is involved in complex activities (e.g. learning and memory). Each hemisphere usually controls actions on the opposite side of the body from which it is situated. The cerebellum is in the midbrain, partly beneath the cerebrum and above the medulla oblongata. In mammals, it is deeply folded and concerned with muscular coordination and regulation of muscle movements. It is also involved in the maintenance of muscle tone and posture. The medulla oblongata forms the hind region of the brain, just above the spinal cord. It contains areas of grey matter responsible for controlling basic functions such as breathing rate, heart rate and some other involuntary functions. The hypothalamus is located on the *ventral* surface of the cerebrum. It contains thermoreceptors for monitoring body core temperature and areas or centres concerned with emotions, *osmoregulation*, sleep, feeding, drinking and speech. It also acts as an *endocrine gland* producing, for example, *hormones* that affect the *pituitary* gland.

breast bone See *sternum*.

breast-feeding Feeding a baby with milk from the breast. All mammals can breast-feed, but humans have the option of bottle-feeding babies with specially formulated milk (cow's milk is unsuitable because it damages the kidneys of newborn babies). Breast-feeding, however, has certain advantages. During the first two days of lactation, a milk-like substance called *colostrum* is secreted, which contains substances that provide the baby with some immunity to infection. Breast milk also contains special fats (called long-chain polyunsaturates or LCPs) which are thought to enhance the development of nervous tissue in the brain.

breathing See *ventilation*.

breeding true See *pure breeding*.

brewing The production of alcohol (e.g. beer and wine) by the fermentation of sugars by *yeast*.

bronchiole See *lung*.

bronchitis See *lung*.

bronchus (*pl.* bronchi) See *lung*.

Brunner's glands See *duodenum*.

bryophyte A member of the plant phylum (or division) Bryophyta that includes the liverworts and mosses. Bryophytes exhibit *alternation of generations*, with the gametophyte generation being the most persistent and conspicuous stage. They do not possess true roots, their bodies being anchored by filamentous rhizoids (delicate hair-like outgrowths). They usually live in damp places and require water for fertilization.

buccal cavity The cavity immediately behind the mouth in the *alimentary canal*, leading to the pharynx and oesophagus. In mammals, it contains the tongue, teeth, and salivary glands and plays an important rôle in the mechanical digestion of food.

budding A form of *asexual reproduction,* found in e.g. hydra or yeast, in which new individuals are formed from an outgrowth of the parent which becomes detached to form a new individual.

buffer A solution that tends to resist changes in *pH* when *acid* or *alkali* is added to it. Buffers usually contain a mixture of a weak acid and its soluble *salt*.

bulb A modified plant shoot (such as that of an onion) that has fleshy leaves which store food. Bulbs are organs of *perennation* and *vegetative reproduction*.

caecum A blind-ending sac in the *alimentary canal* which, in mammals, occurs at the junction of the *small intestine* and *large intestine*. In herbivores, it contains symbiotic bacteria that digest *cellulose*, and may be very large. See also *appendix*.

caffeine A bitter-tasting purine derivative found in chocolate bars, coffee, tea and cola-type drinks. In humans, it acts as a stimulant and *diuretic* (i.e. it increases water losses by stimulating the production of a watery urine).

calciferol See *vitamin D*.

calcium A mineral element essential to living organisms. In plants it is needed to form the middle lamella of *cell walls*. In humans it is the principal component of bone and *enamel* and is also essential for muscular contractions, nerve transmissions, and blood clotting. Deficiency in plants causes stunted growth. Deficiency in humans can cause convulsions and disrupt the normal mineralization of bones; this may lead to poor skeletal growth and rickets in children, or brittle bones in adults. Common sources of calcium are milk and 'hard' drinking water.

calorie The heat required to raise the temperature of 1 gram of water by 1 °C. The Calorie, in which the energy value of food was formerly measured, is 1000 calories. The calorie has been replaced as a unit of work and energy by the joule (1 cal = 4.2 J).

calorimeter See *calorimetry*.

calorimetry The measurement of energy utilized by an organism, usually as heat production measured in joules. Direct calorimetry is the measurement of heat produced by an organism within a chamber, called a calorimeter, supplied with air and surrounded by a jacket of circulating water inside an insulatory layer; heat production is calculated from the increased temperature of the surrounding water. Indirect calorimetry estimates energy utilization from the oxygen consumed by an organism.

Calvin cycle The metabolic pathway (discovered by Melvin Calvin) in the *light-independent stage* of *photosynthesis* in which carbon dioxide is incorporated into more complex chemicals in the plant (see *fixation*) and reduced into carbohydrates using the hydrogen (from NADPH) and *ATP*, both produced during the *light-dependent stage*. The Calvin cycle takes place in the *stroma* of *chloroplasts*. During each turn of the cycle, one molecule of carbon dioxide is fixed into two molecules of *glycerate 3-phosphate* (GP) which is then reduced into two molecules of *glyceraldehyde 3-phosphate* (GALP); three turns are required to release one molecule of GALP for glucose synthesis, leaving five GALP remaining in the cycle to regenerate *ribulose biphosphate*. Since GALP is a three-carbon compound, a total of six turns of the Calvin cycle are required to make one molecule of glucose.

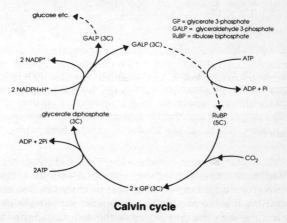

Calvin cycle

calyptra See *root*.

cambium A layer of undifferentiated, thin-walled cells that are situated between the *xylem* and *phloem* in the *vascular* *bundles* of the roots and stems of dicotyledonous plants (see *Dicotyledonae*). Cambium cells can divide by *mitosis* to form new cells which form secondary xylem and secondary phloem. This can continue throughout the life of woody plants, increasing the girth of the stem (see *secondary thickening*).

Canadian pond weed See *Elodea*.

cancer A group of diseases characterized by the uncontrolled proliferation of cells, causing a solid, malignant tumour or other abnormal condition. Cancer cells have the abnormal ability of being able to multiply indefinitely and may invade other tissues. A cancer is usually fatal if untreated. Probably many factors contribute to the development of cancer. Known causative agents include various chemicals (e.g. those in wood smoke and tobacco smoke), ionizing radiation (e.g. X- and gamma-radiation), certain dust particles (e.g. of coal or asbestos), and some *viruses*.

canine tooth One of four sharp, pointed teeth between the *incisors* and premolars of both jaws, in mammals. Canine teeth are particularly prominent in Carnivora, an order which includes dogs, cats and bears, and are used for stabbing and killing prey. Canines are absent in some herbivores (e.g. rabbits).

capillarity or **capillary action** The spontaneous creeping movement of a liquid in very fine tubes or channels, due to *surface tension*. It is the process by which soil water moves in any direction through the fine pores of the soil. Capillarity may make a contribution to the movement of water up *xylem* vessels during *transpiration*, but the effect is small and cannot account for transpiration in large trees; in even the finest xylem vessels, water would rise only about 1 m by capillarity.

capillary The narrowest blood vessel in vertebrates, located between *arteries* and *veins*. Capillary walls are only one cell thick, allowing relatively easy exchange of gases, nutrients and waste products between the blood and tissue fluid. Capillaries can be dilated or constricted to regulate the blood flow through particular tissues.

capillary action See *capillarity*.

carbohydrase An *enzyme* that accelerates the *hydrolysis* of *carbohydrates*.

carbohydrate A member of a group of organic compounds composed of carbon, hydrogen and oxygen, all sharing the chemical formula $C_x(H_2O)_y$. They are major sources of energy, each gram of carbohydrate yielding approximately 16 J (4 calories) of energy when fully respired (see *respiration*). Carbohydrates are classified according to size as monosaccharides (simple sugars), disaccharides, and polysaccharides (including *cellulose*, *glycogen* and *starch*). The types of carbohydrate are described in more detail under their own headings.

carbon The element which is the basis of life on Earth. A carbon atom has the ability to form four covalent bonds oriented in such a way that carbon, hydrogen, oxygen or other atoms can attach to a single carbon atom in any of four different directions. In this way very elaborate three-dimensional structures can be constructed, including straight chains, branched chains, ring-structures, or any combination of these. Consequently, carbon compounds have a unique variety and complexity. Many biologists believe that life as we know it could have evolved only in the presence of carbon.

carbon cycle The cyclical series of processes by which *carbon* atoms are circulated through the *biosphere*. See diagram overleaf.

carbon dioxide

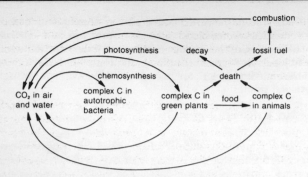

carbon cycle The main steps.

carbon dioxide A colourless, odourless gas, heavier than air, that is a raw material for photosynthesis and is eliminated as a waste product of *aerobic respiration* (and *anaerobic respiration* of plants and *yeast*). It combines with water to form carbonic acid. It comprises about 0.04% of the atmosphere by volume. There is evidence that the atmospheric concentration has increased as result of extensive burning of *fossil fuels* and it is argued that this may result in climatic changes due to the *greenhouse effect*.

carbon monoxide A colourless, odourless gas that is poisonous to mammals; it combines irreversibly with *haemoglobin*, preventing it from carrying oxygen.

carboxyl group The —COOH group found in aldehydes, ketones and many organic acids; it occurs at one end of a *fatty acid* chain.

cardiac cycle The sequence of events that takes place during one heart beat, comprising four main stages: (a) atrial systole (contraction of the atria (see *atrium*) to force blood into the

50

ventricles); (b) a pause; (c) ventricular systole (contraction of the ventricles to force blood through the arteries); and (d) diastole (relaxation of the *cardiac muscle* so that the heart can refill with blood). In a typical adult man, the complete cardiac cycle lasts about 0.8 secs.

cardiac muscle or **heart muscle** Specialized muscle of which the vertebrate heart is composed. Cardiac muscle never fatigues, but it can respire only aerobically and therefore requires a continuous supply of oxygen. Its contractions are myogenic (i.e. they originate from within the muscle), so as long as cardiac muscle is in a solution containing oxygen, an energy source, and the right mixture of mineral salts, it will contract spontaneously without nervous or hormonal stimulation.

carnassial teeth The fourth premolar of the upper jaw and first molar of the lower jaw, in many carnivorous mammals. Carnassial teeth have sharp, overlapping, cutting edges adapted for shearing flesh in a scissor-like manner.

carnivore Any animal that feeds mainly on animal flesh or, more precisely, a member of the order Carnivora, comprising mainly flesh-eating mammals (e.g. wolves, tigers and bears). All Carnivora have large *canine* and *carnassial* teeth to grip and tear flesh, but not all feed exclusively on meat (for example, bamboo shoots are a very important part of the diet of pandas, and many bears feed largely on plant matter).

carpel See *flower*.

carrier 1. A heterozygous individual (see *heterozygote*) that carries a *recessive* allele (see *gene*) for a characteristic which may be harmful. The characteristic (for example haemophilia) is not expressed in the *phenotype*, because the *dominant* allele masks the recessive allele.

2. An individual plant or animal that is infected with a disease-causing organism but shows no signs of the disease. The disease-free carrier can, nevertheless, transmit the disease to other individuals.

carrier molecule A molecule that transports other molecules through biological membranes. Carrier molecules include *proteins* in *plasma membranes* that carry out *facilitated diffusion* and *active transport*.

carrying capacity The greatest number of organisms a particular environment can support during the harshest part of the year, without damage to that environment.

cartilage or **gristle** Tough, flexible, *connective tissue* found in vertebrates. It consists of cartilage cells (*chondrocytes*) surrounded by a firm matrix of *chondrin* (a white, translucent material). Cartilage forms the entire skeleton of all vertebrate embryos and of some fish (e.g. sharks). In other vertebrates, however, most cartilage is replaced by bone during development. In adults, cartilage is confined to certain areas, such as the ends of bones, in joints, and the ears.

catabolic reaction A reaction in a living organism resulting from the breakdown of complex molecules into simple ones (e.g. during *respiration*). Most catabolic reactions liberate *free energy*, but some that the cell performs in order to eliminate unwanted substances actually require free energy. The sum of all the catabolic reactions in an organism is called catabolism. Compare *anabolic reaction*.

catalase An iron-containing *enzyme* that catalyses the breakdown of hydrogen peroxide into water and oxygen:

$$2H_2O_2 \rightarrow 2H_2O + O_2$$

Hydrogen peroxide is a toxic by-product of certain cell reactions, so its elimination is vital to all cells. Catalase is especially abundant in liver and in potato tubers.

catalyst See *enzyme*.

caudal Pertaining to the tail region of an animal. For example, the caudal fin of a fish is its tail fin.

cell The basic structural unit of living matter. Typically, each cell consists of *cytoplasm* bounded by a membrane (see *plasma membrane*) and usually contains *DNA*. In *eukaryotes*, the DNA is organized into *chromosomes* contained within a *nucleus*. In *prokaryotes*, the DNA takes the form of a circular molecule and *plasmids*, which are not contained within a nucleus. See diagram overleaf.

cell body The enlarged region of a nerve cell (*neurone*) that contains the nucleus.

cell capsule A relatively thick and compact layer on the surface of some *bacteria*. If diffuse, it is called a slime layer. The capsule may offer the bacterium some protection and in some cases may link bacteria together into chains.

cell concept See *cell theory*.

cell cycle The sequence of events that occurs in cells of tissues that are actively dividing. It includes a period of normal metabolism (see *interphase*), a period in which the *nucleus* divides (see *mitosis*), and a period in which the cell completes its division to form two new cells (see *cytokinesis*). In very actively dividing tissues the cycle can be completed in less than 24 hours. In cells of a mature organism, the cycle may be slowed down or stopped completely in some tissues.

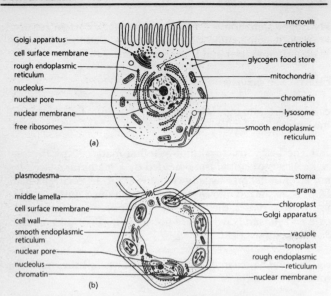

Golgi apparatus
cell surface membrane
rough endoplasmic reticulum
nucleolus
nuclear pore
nuclear membrane
free ribosomes
(a)

microvilli
centrioles
glycogen food store
mitochondria
chromatin
lysosome
smooth endoplasmic reticulum

plasmodesma
middle lamella
cell surface membrane
cell wall
smooth endoplasmic reticulum
nuclear pore
nucleolus
chromatin
(b)

stoma
grana
chloroplast
Golgi apparatus
vacuole
tonoplast
rough endoplasmic reticulum
nuclear membrane

cell A typical animal cell (above) and a typical plant cell (below) as seen under the electron microscope.

cell division The division of a *cell* into two new cells during repair or growth of tissues, or the reproduction of organisms. In *eukaryotes* (the group that includes all animals and plants), cell division occurs as a result of *mitosis* or *meiosis* followed by *cytokinesis*.

cell fractionation See *subcellular fractionation*.

cell membrane One of the *selectively permeable membranes* found both on the surface and inside cells, which consist mainly of *phospholipids* and *protein* (for structure, see *fluid mosaic* *model*). The cell-surface membrane (plasma membrane) acts as a boundary layer, protecting the cell contents. It separates the inside of the cell from its immediate surroundings; these may be another cell surface membrane, tissue fluid, or the external environment. It also acts as a selective mechanism for the movement of materials into and out of the cell. Internal membranes compartmentalize the *cytoplasm*, so that different processes which would otherwise be incompatible can take place simultaneously. These membranes include the double membranes of chloroplasts, *mitochondria*, and the nuclear membrane; the single membrane around *lysosomes;* and the single membranes lining interconnecting system channels called the *endoplasmic reticulum.* The *Golgi body* is thought to make new membranes as the cell grows. Proteins in the membrane have a variety of functions including *active transport, facilitated diffusion*, and the recognition of foreign cells. Some also act as **enzymes**.

cellobiose See *cellulase*.

cell plate A thin partition made from carbohydrate-fused vesicles that is laid down across the middle of a plant cell during *cytokinesis.* The contents of the cell plate become the cell walls between the two cells, and the membranes on either side of the cell plate become the cell surface membranes of the cells.

cell surface membrane See *cell membrane*.

cell theory or **cell concept** The theory that all plants and animals are made of *cells* which have a similar structure. It was first proposed by Matthias Schleiden and Theodor Schwann in 1838—9. In 1858, Rudolph Virchow expanded the theory, emphasizing that all cells arise from pre-existing cells and suggesting that the cell is the basic metabolic, as well as structural, unit of living things.

cellulase An *enzyme* that catalyses the *hydrolysis* of *cellulose* into *cellobiose* (a disaccharide consisting of two β-glucose units) or *glucose*. Relatively few organisms produce cellulase. Mammalian herbivores depend on symbiotic gut bacteria to break down cellulose for them.

cellulose An insoluble *carbohydrate* belonging to the polysaccharides and composed of many *glucose* molecules linked to form straight chains. It is the main structural material of plants and the most abundant organic compound in the world: plants produce approximately 10^{14} kg of it per year, which is roughly equivalent to 70 kg of cellulose per person per day. Most cellulose is contained within plant *cell walls* where groups of about 2000 cellulose molecules form interweaving microfibrils which give the cell wall its strength. The linear arrangement of cellulose allows strong *hydrogen bonds* to form between adjacent molecules, strengthening cellulose and making it very resistant to *hydrolysis*. Only *bacteria*, *fungi* and protozoa can make their own cellulose-digesting enzyme, *cellulase*.

cellulose

cell wall A relatively rigid wall surrounding the cell surface membrane of plant cells, *fungi* and *prokaryotes*. In plants it is formed of *cellulose* fibres embedded in a matrix consisting of *polysaccharides* and *proteins*. Fungal cell walls are composed mainly of *chitin*, those of prokaryotes are made of murein, a complex made of polysaccharides and *amino acids*. Plant cell walls help to protect and maintain the shape of the cell and prevent it from bursting when immersed in pure water. Most plant cell walls are completely permeable to water, but in some plant tissues (e.g. *xylem*) they may be strengthened and made impermeable by the addition of *lignin*.

central nervous system In vertebrates, the part of the nervous system comprising the *brain* and *spinal cord*. It lies between the receptors and *effectors* and acts as an integration centre, processing and relaying information.

centrifuge A machine that spins substances contained within tubes so the centrifugal forces generated separate molecules from solution, particles and solids from liquids, and immiscible fluids from one another. The separation depends on the different densities of the substances. See also *ultracentrifugation*.

centriole An *organelle*, occurring in animal cells but absent from most plant cells, which comprises a hollow cylinder composed of nine groups of triplet *microtubules* held together by proteins. Centrioles usually occur in pairs at right angles to each other near the nucleus. They pull apart during *mitosis* and *meiosis* and are believed to be involved in the organization of the *spindle apparatus*. Centroles have their own DNA and may be self-replicating.

centromere A constricted part of a *chromosome* which contains no *genes*; it is the region in which sister *chromatids* are joined. Centromeres produce a complex system of fibres which attach to the *spindle apparatus* and facilitate the separation of chromosomes and sister chromatids during *mitosis* and the *anaphase* of *meiosis*. The centromere duplicates during the metaphase of mitosis and the second division of meiosis, so sister chromatids can be moved separately.

centrum See *vertebra*.

cephalization The development of the head in the course of animal evolution. It arises mainly because animals are motile and bilaterally symmetrical and tend to move with one particular part of their body in front. A concentration of sense organs and nervous tissue at the anterior (front) end enables them to detect the environmental conditions into which they are moving.

Cephalopoda See *Mollusca*.

cerebellum See *brain*.

cerebrum See *brain*.

cervix A narrow, neck-like part of an organ. In mammals, the name usually refers to the region lying between the *uterus* and the *vagina*, whic h contains a ring of muscle and mucus-secreting cells.

chaetae See *Annelida*.

character or **trait** Any detectable attribute or property of an organism which forms part of its *phenotype*.

checkerboard See *Punnett square*.

cheese-making An example of *biotechnology* in which microorganisms are used to ferment milk to form a coagulated product. Different cheeses are produced by the use of different microorganisms and types of milk (e.g. from cow, sheep or goat).

chemical bond energy A term sometimes used to describe the bonds linking the three phosphate groups to adenosine in *adenosine triphosphate* (ATP). These bonds can undergo *hydrolysis* to release a large amount of energy, so they are sometimes described as high-energy bonds, but this is incorrect. When ATP is hydrolysed, not only is phosphate removed but water is added; bonds are formed as well as broken. The energy released is a property of the whole reaction between ATP and water.

chemical control See *pest*.

chemical coordination The use of chemicals to enable different parts of an organism to work together. See *endocrine system* and *plant growth substance*.

chemoautotroph or **chemosynthetic organism** An organism that uses carbon dioxide as a starting material for the synthesis of complex organic molecules. The energy for the synthesis comes from oxidation of inorganic substances rather than from light. Nitrifying bacteria are chemoautotrophs that play an important role in maintaining soil fertility through their activities in the *nitrogen cycle*. See *autotrophic nutrition*.

chemosynthesis The synthesis of complex organic compounds using carbon dioxide as a carbon source and energy from chemical reactions rather than light. See *chemoautotroph*.

chemosynthetic organism See *chemoautotroph*.

chemotaxis See *taxis*.

chemotropism See *tropism*.

chewing the cud Rumination. See *rumen*.

chi-squared test or **χ^2 test** A statistical procedure for testing the significance of the deviation between numbers observed in an experiment or investigation and numbers expected from a given hypothesis. The measure of the deviation is called the chi-squared value and is calculated using the following equation:

$$\chi^2 = \Sigma(d^2/x)$$

where χ^2 is chi squared, d is the difference between the observed and expected results (i.e. the deviation), and x is the expected result.

chiasma (*pl.* **chiasmata**) A point at which *homologous chromosomes* make contact during *prophase* of the first division of *meiosis*. They are the points at which *crossing-over* occurs. Several chiasmata may form in a *bivalent*.

chitin A structural polysaccharide which is the main component of *arthropod cuticles*. It is also found in fungal *cell walls*. Chitin molecules are usually arranged in layers which may be cross-linked to yield a very strong, lightweight material. Chitin is similar to *cellulose*. Both are made from β-glucose units, but in chitin one of the —OH groups in each *glucose* molecule is replaced by an —NH.CO.CH$_3$ group. Chemically, chitin is a glycosamine polymer.

Chlorella A *genus* of microscopic, green, non-motile, *unicellular organisms*. Each cell usually has a single, cup-shaped *chloroplast* and feeds only by *photosynthesis*. *Chlorella* reproduces asexually, cells dividing internally four times to form 16 daughter cells initially enclosed within the parent cell wall. *Chlorella* is common in many aquatic and terrestrial habitats, especially small bodies of standing water.

cholecystokinin See *pancreomyzin*.

chlorophyll A group of green pigments found in the chloroplasts of most plants and on special membranes in cyanobacteria (formerly known as blue-green algae; see *bacterium*). Chlorophyll consists of a porphyrin ring with magnesium at the centre and linked to a long-chain alcohol (phytol). Chlorophyll *a*, the only type common to all plants and the only one found in cyanobacteria, absorbs red and blue light, using the energy to drive the reactions of photosynthesis. Photosynthetic bacteria have other kinds of chlorophyll, called *bacteriochlorophyll*, which contain manganese instead of magnesium.

chloroplast A double-membraned *organelle* found in plants and some unicellular organisms. Plant chloroplasts are typically lens-shaped and about 5 μm long. They contain *chlorophyll* embedded in stacks of membrane-filled sacs (called *grana*) which are involved in the *light-dependent stage* of *photosynthesis*. The gel-like matrix of chloroplasts is called the *stroma* and contains enzymes required for the *light-independent stage*. It also contains *DNA*, *ribosomes*, phosphate granules, oil droplets and *starch granules*.

chlorosis See *magnesium*.

cholesterol A *lipid* found in the *cell membranes* of most *eukaryotes*; it helps strengthen the membranes. Cholesterol is a precursor of a number of substances, including steroid *hormones*, *bile* acids, and *vitamin D*, and can be synthesized in the *liver* from *acetyl coenzyme A*. In humans, excess cholesterol may be carried by the blood and deposited in arterial walls, narrowing the *arteries* and increasing *blood pressure*. High blood cholesterol is believed to increase the risk of heart disease.

Chondrichthyes A class of fish in which the skeleton is made entirely of *cartilage*. The mouth is ventral (on the underside of the head), fins are fleshy, and the gill openings are separate with no *operculum*. Compare *Osteichthyes*.

chondrin See *cartilage*.

chondrocytes See *cartilage*.

Chordata or **chordates** A phylum of animals which includes fish, amphibians, reptiles, birds and mammals. At some stage in their lives they have visceral clefts (openings in the wall of pharynx which may function as gill slits), a *notochord*, and a hollow, dorsal nerve cord. In vertebrates, the notochord develops into the backbone which surrounds and protects the spinal cord. All adult chordates have a tail behind the *anus*.

chordates See *Chordata*.

chorion See *chorionic villus*.

chorionic villus A small, finger-like outgrowth of the outer membrane (*chorion*) enclosing the embryos of reptiles, birds and mammals. In humans, chorionic villi project into blood spaces in the *uterus* of the mother. During early stages of development they contain the rapidly dividing cells of the foetus. These foetal cells can be sampled by inserting a small tube through the *vagina* and *cervix*; the cells can then be analysed for genetic abnormalities. This procedure of genetic screening is called chorionic villus sampling. Its main advantage over *amniocentesis* is that it can be performed as early as the ninth week of pregnancy.

choroid layer See *eye*.

chromatid One of two copies of a *chromosome* which has replicated during *interphase*. The sister chromatids (the pair of chromatids which make up a chromosome) appear as threadlike structures during *prophase* of nuclear division. They are held together at the *centromere* until being separated during *anaphase* of *mitosis* and in the second division of *meiosis*.

chromatin A granular substance present in the *nucleus* of a cell during *interphase*. It consists of a network of chromosomal material (*DNA* and *proteins*) with a little *RNA*. 'Chromatin' means 'coloured material', with reference to the ease with which it becomes stained in the laboratory. Inactive chromatin (called *heterochromatin*) appears dark when stained because the chromosomes are in a more condensed form. Active chromatin (*euchromatin*), in which messenger RNA molecules are formed, is stained more lightly because the chromosomes are in a more dispersed state.

chromatography A technique used to analyse or separate the components of a mixture of gases, liquids, or dissolved substances (e.g. *amino acids* or *chlorophyll* pigments). The technique depends on the different solubility of molecules in a moving solvent (mobile phase) and absorptions on, or solubilities in, an inert substrate such as paper, chalk, or silica gel (stationary phase). Components of the mixture are carried to different parts of the substrate (or pass through the substrate at different rates) and thus separate out. They can be identified by comparison with the movement pattern of known molecules.

chromosome The structure that contains *DNA* and carries genetic information. Each chromosome is composed of thousands of *genes* arranged in a linear sequence. In *eukaryotes*, the DNA is associated with proteins (including *histones*); in *prokaryotes* (bacteria) the chromosome is a loop with no associated proteins. Chromosomes contained within the nucleus of eukaryotic cells are strongly stained by various dyes, including acetic orcein. With the aid of a light microscope, they become visible as rods or thread-like structures during *prophase* of *mitosis* and *meiosis*. In diploid organisms (see *diploidy*), chromosomes exist in homologous pairs, and each species has a typical number (for example, in human cells there are 23 pairs).

chromosome mutation See *mutation*.

chrysalis See *pupa*.

chyme See *duodenum*.

ciliary body A circular band of tissue that surrounds and supports the lens of the vertebrate *eye*. It contains muscles (ciliary muscles) to which the lens is attached and glands which secrete aqueous humour. The ciliary muscles change the shape of the eye and bring about focusing.

ciliary muscle See *ciliary body*.

ciliated epithelium An epithelial layer (see *epithelium*) containing cells with *cilia* (see *cilium*).

cilium (*pl.* **cilia**) A fine, thread-like projection (up to 10 μm long) of the surface membrane of a cell (see *cell membrane*), which moves the fluid surrounding it by a beating or rowing action. Cilia can produce cell movement (as in some *unicellular organisms*) or move material along the surface of the cell. Internally, a cilium has two central *microtubules* and nine sets of doublet microtubules around the circumference (the so-called '9 + 2' structure).

circadian rhythm The rhythm of activity seen in many organisms kept under constant environmental conditions. These circadian rhythms usually have a periodicity slightly longer or shorter than 24 hours. In contrast, organisms in natural habitats usually exhibit diurnal rhythms, strictly 24 hours in length, synchronized to the daily light—dark cycles.

circulation Movement of a fluid within a *circulatory system* in an organism.

circulatory system A system that transports materials around the body in a fluid (blood), in an animal that is too large for transport by *diffusion* to be effective. Circulatory systems may be open or closed, single or double. In an open circulatory system, blood moves freely within body spaces for much of its circulation (e.g. in arthropods, the blood lies in the main body cavity, the *haemocoel*, and bathes the tissues). In closed circulatory systems, the blood is confined within *blood vessels*: in vertebrates, the closed blood circulatory system consists of heart, arteries, capillaries and veins. Blood may pass through the heart once during every complete circuit of the body (single circulatory system, e.g. of fish) or twice during each circuit (double circulatory system, e.g. of mammals). See also *vascular system*.

cisternum (*pl.* **cisternae**) A fluid-filled compartment, usually flattened, formed within the *cytoplasm* of a cell by membranes of the *endoplasmic reticulum* and *Golgi body*.

class See *taxon*.

cistron A portion of DNA that contains the code for a single polypeptide chain. It is sometimes called a *functional gene*.

classification (in biology) The arrangement of organisms into groups. Classifications can be made either on the basis of one or a few characteristics selected for convenience, such as colour, form of locomotion, or habitat (artificial classification), or on the basis of characteristics which reflect an evolutionary relationship between the organisms grouped together (natural classification), such that members of the same group have a common ancestor.

clavicle (in mammals) One of a pair of slender, doubly curved, long bones extending horizontally across the top of the chest, forming the anterior or ventral part of the shoulder girdle. In humans, the clavicle forms the collar bone and is firmly attached to the *scapula* (shoulder blade) and *sternum* (breastbone).

climate The long-term atmospheric conditions that characterize a particular locality, based on a suitably long period of weather records (e.g. 30 years). Factors considered include variations in temperature, humidity, rainfall, cloud cover, wind direction and force, and atmospheric pressure.

climax A community of organisms that forms the final stage in natural *succession;* the climax community has reached an approximately steady state and is in equilibrium with its environment. The natural climax in most of lowland Britain is oak woodland. Succession does not always proceed naturally. Human interference may prevent the natural climax from occurring. For example, much heathland is maintained by human activities, such as burning and grazing, without which the area would be colonized by scrub and develop into woodland; heathland is thus said to result from deflected succession and is called a *deflected climax* or *plagioclimax*.

clitoris Erectile tissue lying in front of the *urethra* in female mammals. It is homologous to the male penis.

cloaca A common chamber in the pelvic region, with a single opening to the exterior, into which the contents of the *alimentary canal*, *kidneys*, and reproductive organs discharge. It is found in most vertebrates, but not in mammals apart from the platypus and the echidnas (spiny anteaters).

clone 1. A group of genetically identical cells or organisms arising from *mitosis*.
2. An identical copy of a particular strand of DNA.

closed circulatory system See *circulatory system*.

Cnidaria or **coelenterates** A phylum of animals in which the body comprises two layers of cells separated by a jelly-like non-cellular layer (called *mesoglea*). The body is typically radially symmetrical (see *radial symmetry*) and possesses tentacles armed with *nematoblasts* (also called *cnidoblasts*). These are cells containing a long coiled thread (the *nematocyst*) which is discharged on contact with prey. Nematocysts may inject a poisonous substance into the prey or entwine the prey and immobolize it. Jellyfish (e.g. *Aurelia*), *Hydra* and sea anemones are members of the Cnidaria.

cnidoblast See *nematoblast*. See also *Cnidaria*.

coagulation The change from a liquid to a viscous or solid state by the clumping together of colloidal particles (see *colloid*). It may be caused by heating or by a chemical reaction. See also *blood clotting*.

cobalamin See *vitamin B12*.

coccus (*pl*. **cocci**) A *bacterium* that has a roughly spherical shape. Cocci which cause pneumonia (*Diplococcus pneumoniae*) occur in pairs. Staphylococci, some strains of which cause boils, occur in bunches like grapes.

cochlea A fluid-filled tube in the inner ear, consisting of three parallel canals, spirally coiled like the shell of a snail, and containing sensory cells located in the *organ of Corti*. The cells respond to movements of fluid in the cochlea caused by the transmission of sound vibrations from the outer ear, different points along the length of cochlea responding to different sound frequencies.

cocoon See *pupa*.

codominance The condition which occurs in a heterozygote, when both *alleles* of a particular pair influence the *phenotype* of the individual. In genetics problems, where codominance exists (e.g. in the inheritance of sickle cell anaemia, and the human ABO blood group) the *gene* is designated by an upper case (capital) letter and the alleles by appropriate superscript letters. See *dominance*.

codon or **triplet** A group of three nucleotide bases in a *nucleic acid*, that code for a specific *amino acid* or act as a signal to start or stop the expression of a *gene*. By convention, the term codon generally refers to the triplets in messenger RNA (e.g. UUU specifying phenylalanine, and UAA, UAG, and UGA being codons which terminate translation). See also *genetic code*.

coelenterates See *Cnidaria*.

coelom A fluid-filled cavity formed by the splitting of the middle layer of cells (*mesoderm*) in animals with three body layers. The coelom is the second and main body cavity, separating the muscles of the gut (the first body cavity) from those of the body wall, thus allowing them to move independently of each other, and providing an area for the enlargement of internal organs, permitting the gut to become differentiated for various functions. It also plays an important rôle in collecting waste products and can act as a storage site for the maturation of *gametes*.

coelomate (of an animal) having a *coelom*.

coenzyme See *cofactor* and *acetyl coenzyme A*.

cofactor A non-protein substance essential for the normal activity of an *enzyme*. Cofactors may be inorganic ions (*activators*) or organic molecules (*coenzymes*). Activators (often metal ions) enable *substrates* to combine with enzymes more easily.

Iron, for example, is the activator required for *catalase*. Coenzymes usually take part in the enzyme-catalysed reactions by donating or accepting certain chemical groups, but they are regenerated in their original form at the end of the reaction. *NAD* and *ATP* are two common coenzymes. Many coenzymes are derived from *vitamins* (e.g. NAD is derived from nicotinic acid, a member of the vitamin B complex).

cohesion The force of attraction between molecules of the same substance (e.g. the *hydrogen bonding* by which water molecules tend to stick together). See also *transpiration*.

coleoptile A conical, protective sheath that surrounds the plumule (embryonic shoot) in a germinating monocotyledonous seed such as grass (see *Monocotyledonae*). It is the first structure to emerge above ground.

collagen An insoluble, fibrous *protein* that accounts for 30% of the total body protein of mammals. It is an important component of *connective tissue* in skin, bones and tendons. The polypeptide chains of collagen form triple-stranded helical coils (see *alpha helix*), bound together to form fibrils of great tensile strength but limited elasticity.

collar bone See *clavicle*.

collecting duct A tube or duct that conveys urine from the *kidney* cortex to the *ureters*.

collenchyma A plant tissue composed of elongated living cells in which the primary cell wall is unevenly thickened with *cellulose* deposited at the corners. It is specialized to provide mechanical support to actively growing parts of the plant, especially the young stems and midribs of leaves, which may also need to be flexible.

colloid A substance (e.g. *protein*) whose particles range in size from 1 to 100 μm. The particles do not go into solution easily and do not settle out under gravity. When mixed with water colloids tend to appear opaque rather than clear. They have a high capacity for absorbing water and other substances and therefore play a very important role in holding molecules in position in cells and maintaining organization within the protoplasm. They do not pass easily through cell membranes.

colon The first part of the *large intestine*, containing undigested food from the *small intestine*. The main function of the colon is to absorb large amounts of water from the undigested food and to form the *faeces*, which are passed on to the *rectum*. The colon also secretes mucus and may absorb some vitamins and minerals.

colorimetry A technique for measuring the concentration of a coloured solution using a device called a *colorimeter*. Light passes from a bulb via an appropriate filter (this makes the calorimeter more sensitive) through the sample and onto a photosensitive cell. The deeper the colour of the solution, the smaller the amount of light transmitted and falling onto the cell. Colorimeters show either the amount of light transmitted or the amount absorbed (or both). Colorimeter readings from test samples are compared with readings from solutions of known concentration.

colostrum See *breast-feeding*.

colour blindness A genetic disorder in humans, resulting in an inability to distinguish between pairs of colours, usually red and green, although the ability to distinguish shade is unaffected. The *recessive* allele for colour blindness is a form of a gene carried on the X chromosome, therefore colour blindness is a sex-linked characteristic more common in males than females.

colour vision See *vision*.

column graph A graph that shows the frequency distributions of discontinuous or discrete data. See appendix 1.

commensalism An association between two animals of different species which live together and share food resources. One partner benefits, but the other is neither harmed nor gains benefit. See also *mutualism*.

communication The transfer of information from one part of an organism to another, or from one organism to another. Within an organism, information is conveyed by nerve impulses or chemicals (see *hormone* and *plant growth substance*). Between organisms, information may be conveyed through any of the sense organs. Many organisms communicate by visual displays and sound.

community A localized grouping of a number of populations of different species. A community is the living component of an *ecosystem*, its members interacting to form relationships (especially feeding relationships) which make the community relatively self-contained.

companion cell A small, thin-walled cell in plants, situated in the *phloem*. Companion cells have nuclei and are densely packed with *organelles*. Each cell lies next to a sieve element to which it is connected by cytoplasmic threads (*plasmodesmata*) passing through the cells walls. The companion cell appears to regulate the activity of the sieve tube and plays an essential role in **translocation**.

compensation point The point at which the rate of *photosynthesis* in a plant is in exact balance with the rate of *respiration*, so there is no net exchange of carbon dioxide or oxygen, due either to the low intensity of light or the low carbon dioxide concentration to which the plant is exposed.

competition Interaction between organisms that are striving for the same environmental resources (e.g. food, space, or a mate). The organisms may belong to the same species (*intraspecific competition*) or different species (*interspecific competition*). Competition tends to affect adversely the growth rate, survival, or reproductive output of one or both competitors. Competition in which one organism gains the whole resource (e.g. a mate) is called a *contest competition*. Competition in which all organisms obtain less of a scarce resource (e.g. food or living space) is called *scramble competition*. Where a resource is very limited, scramble competition may be so intense that some organisms die.

competitive inhibition See *inhibition*.

compost 1. A soil conditioner made by mixing organic waste (e.g. rotted plant material and animal dung) in a container in order to accelerate decomposition.
2. A medium in which plants (especially potted plants) are cultivated.

compound 1. A substance composed of two or more elements in definite proportions by weight.
2. (of biological structures) Made up of several similar parts (e.g. compound leaves consist of several leaflets).

concentration gradient The difference between the concentration of molecules or ions at two locations (e.g. on the outer and inner surfaces of a *cell membrane*).

conception (in mammals) The *fertilization* of an egg cell by a *spermatozoon* in the *uterus*.

condensation A reaction involving the removal of water from two or more molecules so that they can combine to form a larger molecule. *Amino acids* combine by condensation to form polypeptide chains of *proteins*. Compare *hydrolysis*.

conditional stimulus See *reflex action*.

conditioned reflex See *reflex action*.

cone 1. A a cone- or flask-shaped sensory cell in the *retina* of vertebrate eyes that is responsible for colour vision and detailed vision in bright light. Individual cones seem to be sensitive to blue, green, or red wavelengths and are mainly concentrated in the *fovea* of the retina.
2. A plant reproductive structure. See *Coniferophyta*.

confidence interval A range of values so determined by a statistical method as to have a prescribed probability of containing the true value of an unknown parameter.

Coniferae See *Coniferophyta*.

Coniferophyta or **conifers** A group of vascular plants (tracheophytes) which bear cones, but do not produce flowers or fruit. Cones are reproductive structures that consist of a conical mass of spore-producing leaves (usually hardened). In other classifications, the Coniferophyta are ranked as the subdivision Gymnospermae and the coniferous trees as the class Coniferae.

conifers See *Coniferophyta*.

conjugated protein A molecule, such as haemoglobin, consisting of a protein combined with a distinct non-protein group.

conjunctiva A thin layer of mucus-secreting *epithelium* and associated *connective tissue* that covers and protects the front of the vertebrate *eye* and lines the eyelids.

connective tissue Animal tissue in which the material between the cells (the *intercellular matrix*) forms a major part. There are many types of connective tissue with widely differing properties and functions. Loose connective tissue (also known as *areolar tissue*) binds together organs within the body cavity, and binds the skin to underlying structures. Its transparent matrix contains flexible fibres, some elastic, some inelastic, which give this tissue great strength and resilience. *Adipose tissue* stores fat. Bone and *cartilage* protect enclosed structures and provide mechanical support. *Blood* is a fluid connective tissue with a wide range of functions, including transport of materials.

conservation Protection of something desirable against change, loss or injury. In biology, this is generally nature conservation, involving the active management of *ecosystems* to retain their quality, value and diversity, while acknowledging human requirements and the naturally dynamic character of ecosystems. Conservation usually involves positive interference (e.g. culling to prevent overpopulation).

contest competition See *competition*.

continuous flow processing See *fermentation*.

continuous variation See *variation*.

contraception The deliberate prevention of *fertilization*. In humans, there are several contraceptive methods. The rhythm method depends on monitoring the *menstrual cycle* so *ovulation* can be pinpointed and sexual intercourse avoided during fertile periods. Barrier methods use physical barriers (e.g. a condom, a thin latex sheath placed over the penis) to prevent *spermatozoa* from reaching the egg cell. Chemical methods include spermicides which kill spermatozoa and contraceptive pills which mimic the effects of sex *hormones*. The combined pill contains *oestrogens* and *progesterones* that prevent preg-nancy by inhibiting ovulation and altering the environment of the *vagina* and *uterus*. Sterilization is the ultimate form of con-traception; it involves cutting or tying the tubes along which sperm or eggs travel: the vas deferens in men and the Fallopian tubes in women.

contractile vacuole An *organelle* that is thought to carry out *osmoregulation* in some *unicellular organisms* (e.g. *Amoeba*). In freshwater ponds, an *Amoeba* would be in danger of becom-ing over-inflated and bursting because of water which enters it by *osmosis*. The contractile vacuole prevents this: salts that are actively pumped (see *active transport*) into the vacuole draw water into it; when full, the vacuole empties its liquid contents into the surrounding medium, maintaining a water balance in the *Amoeba*.

control A procedure to ensure that observations or data obtained from an experiment are due to the factor under inves-tigation and not some other factor. A control provides a stan-dard against which experimental results can be compared. It is performed at the same time as the experiment and usually dif-fers from it in only one factor, the one under investigation.

controlled factors Factors that are varied or held constant by an investigator during an experiment, according to pre-determined specifications.

convoluted tubule See *nephron*.

copulation The insertion of the erect penis of a male mammal into the vagina of a female. Copulation is necessary for internal fertilization to take place. During copulation, spermatozoa are ejaculated from the penis into the female reproductive tract, from where they may find their way to an egg cell and fertilize it.

cork Dead, waterproof plant cells formed from the layer of cells immediately inside the *epidermis*. See also *bark*.

corm A short, enlarged, fleshy, vertical, underground plant-stem (e.g. of *Crocus*) that functions as a perennating organ (see *perennation*) and in *vegetative reproduction*. The stem is surrounded by protective scale leaves (remains of the previous season's foliage leaves) but, unlike the leaves of *bulbs*, these are not swollen.

cornea A disc-shaped, transparent layer of *epithelium* and *connective tissue* in front of the iris and lens of the *eye*. Light is refracted as it passes through the cornea so it begins to converge before reaching the lens. In land-dwelling vertebrates, the cornea provides most of the eye's focusing power and the lens is used for fine focusing of the light on to the retina.

corpus luteum See *Graafian follicle*.

cortex The outermost part or layer found in a variety of plant and animal structures and organs. In plant stems and roots, the cortex lies immediately inside the *epidermis* but outside the *vascular bundles*. It usually consists of mainly undifferentiated *parenchyma* cells.

cotyledon A seed-leaf; part of a plant *embryo* that may be specialized for food storage. In some plants, the cotyledons are pushed above the ground and act as the first photosynthesizing (see *photosynthesis*) leaves.

counter-current exchange A process in which two fluids flow in opposite directions along adjacent channels, thus maximizing the exchange of gases or heat. For example, in the *gills* of bony fish, water is pumped past the gill filaments in the opposite direction to the flow of blood; a *concentration gradient* maintained along the whole length of the gills ensures that gas exchange also takes place along the whole length.

courtship Behaviour, in some animals, that precedes *copulation*. Courtship behaviour is often highly ritualized, involving displays and posturing, and the secretion of chemical attractants. One function of courtship may be selection of a specific mate from a number of potential ones. Another may be to overcome fear or aggression between animals when approached by strangers. In species in which the male and female remain together for a long time to care for their offspring (e.g. humans), courtship is important in establishing a strong bond between partners.

COV See *gene mapping*.

cranium Part of the bony vertebrate skeleton that encloses and protects the fragile brain and organs of hearing and balance. It consists of several flat bones that become fused together in adults.

crista (*pl.* **cristae**) One of the shelf-like folds of the inner membrane of a *mitochondrion*.

crop rotation The growing of different crops in a field in regular sequence over a number of seasons. Crops vary in their nutrient requirements, so rotation reduces the risk of depleting the soil of minerals. Legumes (e.g. peas and clover) are usually included in the rotation to replenish soil nitrogen through the nitrogen-fixing bacteria they contain in their *root nodules*. Rotational cropping also restricts the growth of pest populations peculiar to particular crop species. Compare *monoculture*.

cross-fertilization See *fertilization*.

cross pollination See *pollination*.

crossing-over A natural process, occurring during *prophase* of *meiosis*, in which there is an exchange of genetic material between *homologous chromosomes*. *Chromatids* forming a *bivalent* break at points of contact (*chiasmata*) and rejoin in such a way that sections are exchanged. This exchange of genetic material is also called *recombination*, but recombination can also be brought about by artificial genetic manipulation. Crossing-over leads to increased variation.

cross-over value See *gene mapping*.

crustaceans See *Arthropoda*.

crypts of Lieberkuhn Narrow pits of glandular tissue formed between two adjacent villi in the *small intestine*, through which intestinal secretions pass. The base of the crypts contain cells (*Paneth cells*) that secrete mucus and alkaline salts.

cud See *rumen*.

cultivation 1. The preparation of the ground (e.g. by plough-ing, digging and draining the soil) for growing crops.
2. The planting, care, and harvesting of crops.

culture Cells, tissues or microorganisms being grown on a nutrient medium under artificial, controlled conditions.

cuticle 1. A non-living, waterproof layer on the outer surface of some plant cells, especially the shoots of flowering plants. It is made of a waxy material (cutin) secreted by the *epidermis*. Its main function is to reduce desiccation, but it may also offer some mechanical support and protection against diseases.
2. A protective outer layer that covers the body of many inver-tebrates. Its composition varies. In many *arthropods* the outer-

most part (*epicuticle*) contains lipoprotein and wax. Beneath this are two layers containing *chitin:* the outer layer (*exocuticle*) is usually hardened and the inner (*endocuticle*) remains flexible. In order to grow, arthropods must shed their cuticle. See also *exoskeleton*.

cutin See *cuticle*.

cutting In flowering plants, a method of artificial *vegetative reproduction* from a stem or leaf. The stem (e.g. of a rose plant) is cut just below a *node* and excess leaves removed. The leaf (e.g. of an African violet) is cut off at the leaf base. A cutting is dipped in fungicide and placed in water or moist sand. Once *adventitious roots* (small lateral roots growing out of the stem) have developed, the cutting is transferrred to soil. Sometimes an *auxin* (a *plant growth substance*) is applied to the cut end to encourage growth of the roots.

cyclic AMP or **cyclic adenosine monophosphate** A ring structure that can be formed in the *cytoplasm* of a cell from *adenosine triphosphate* (ATP) by the action of an *enzyme* in the *plasma membrane*. Changes in levels of cyclic AMP affect metabolic reactions by inhibiting or activating enzymes. It acts as an intracellular or second messenger in some reactions induced by *hormones* and is also involved in *cell division*, *gene expression*, immune responses, and nervous transmission.

cyclic photophosphorylation See *light-dependent stage*.

cyclosis or **cytoplasmic streaming** The phenomenon in which the entire *cytoplasm* circulates around a plant cell in the space between the *vacuole* and the surface membrane. Microfilaments made of a contractile protein similar to *actin* are thought to be involved.

cystic fibrosis A human genetic disorder resulting from a *recessive* allele carried on chromosome 7. In *homozygotes*, the recessive allele appears to disrupt the transport of chloride ions into and out of cells, causing a malfunction of the *pancreas* and abnormal secretion of mucus in the lungs which can lead to secondary conditions (e.g. infections). Cystic fibrosis can be fatal in children if left untreated. It is one of the first inherited diseases to be treated by *gene therapy*.

cytokinesis The division of cell *cytoplasm* so that two daughter cells can form after *mitosis* or *meiosis*. In animals, it is synonymous with cell division and involves the cell surface membrane being drawn in at the middle of the cell to form a furrow. In plants, it involves the formation of a *cell plate* and *cell wall* to separate the two daughter cells.

cytokinin See *plant growth substance*.

cytology The scientific study of the structure and function of cells.

cytoplasm The living part of a cell inside the cell surface membrane (see *cell membrane*) and outside the *nucleus*. It consists of a watery fluid, the *cytosol*, in which *organelles* are suspended. A network of microfilaments and *microtubules* (called the *cytoskeleton*) provides support and maintains the shape of the cytoplasm. The cytoskeleton may also be involved in movement of material within the cell, or movement of the cell itself.

cytoplasmic streaming See *cyclosis*.

cytosine See *pyrimidine*.

cytoskeleton See *cytoplasm*.

cytosol The fluid part of *cytoplasm*, which contains numerous small molecules (e.g. sugars, salts, amino acids and nucleotides) in solution, and larger molecules, such as proteins, in colloidal suspension (see *colloid*).

dark stage See *light-independent stage*.

Darwinism The theory of evolution proposed by Charles Darwin (1809—82). This maintains that new species evolve from ancestral types by the action of *natural selection* on *variations* within populations. A modern version of Darwinism (called *neo-Darwinism*) incorporates discoveries about *genes* and genetic change that explain the source of the variations upon which natural selection acts. This was an aspect of evolution which Darwin was unable to explain satisfactorily.

data See *datum*.

datum (*pl.* **data**) A single item of information.

daughter cells New cells (irrespective of sex) that result from *cell division*. Following *mitosis*, two daughter cells are formed; following *meiosis*, four daughter cells are formed.

DCPIP test A test for ascorbic acid. DCPIP (phenolindo-2,6 dichlorophenol) is a poisonous, blue dye which is decolorized by protons (hydrogen ions). Ascorbic acid (*vitamin C*) releases protons when in a solution of water, the number of protons depending on the concentration of the ascorbic acid. DCPIP can therefore be used as an indicator for determining the concentration of vitamin C in various drinks, compared with a standard solution of ascorbic acid of known concentration.

DDT (DichloroDiphenylTrichloroethane) A chlorinated hydrocarbon which was the first major synthetic insecticide.

DDT is cheap and effective but is not easily broken down and tends to accumulate in fatty tissue, resulting in its transfer from one consumer to another up the *food chain*. This process of accumulation in members of a food chain or *food web* of materials present only in minute amounts in the environment is called *bioaccumulation* or *biological magnification*. Bioaccumulation of DDT resulted in top predators (e.g. carnivorous birds) having high concentrations of DDT. The concentrations were not high enough to kill the birds, but they caused eggshells to be weakened, so that fewer offspring survived. Since the 1970s the use of DDT has been banned in many developed countries, but it is still used in some developing countries.

deamination The removal of an amino group (—NH$_2$) from a molecule. In mammals, deamination takes place in the *liver* to break down excess *amino acids*. Toxic ammonia (NH$_3$) is formed from the amino group and enters the *ornithine cycle* to produce less harmful *urea*. The residue of the amino acid can enter the *Krebs cycle* and be used as an energy source in *aerobic respiration*.

death rate See *mortality*.

decarboxylation The removal of a carboxyl group (—COOH) from a molecule, usually resulting in the formation of carbon dioxide (e.g. in the *Krebs cycle*). See also *oxidative decarboxylation*.

decay The decomposition of organic material through the action of microbes.

deciduous Easily shed. Examples of deciduous structures are the scales of some fish and the leaves of a *deciduous plant*.

deciduous plant A plant in which all the leaves are shed at the end of a growing season. In temperate regions, leaf fall is usually in the autumn (or fall) and in tropical regions at the beginning of the dry season. Leaf fall allows the plant to retain water which would otherwise be lost by *transpiration*.

deciduous teeth See *milk teeth*.

decomposer An organism (e.g a fungus or bacterium) that obtains its nutrients by feeding on dead organisms and breaking them down into simpler substances, ultimately releasing inorganic materials. Decomposers play a vital role in *nutrient recycling*.

defecation or **defaecation** The elimination of *faeces* from the *alimentary canal*.

deficiency disease A disease caused by the lack of an essential nutrient.

deflected climax See *climax*.

deforestation The permanent removal of trees and undergrowth to clear large areas of forest and woodland. On steep slopes, this can lead to increased soil erosion and siltation of rivers and lakes. Deforestation may increase carbon dioxide levels in the atmosphere, through the loss of photosynthesizing trees and deterioration of soil structure and increased rates of decomposition in the soil.

degenerate code See *genetic code*.

deglutition See *swallowing*.

dehiscent See *fruit*.

deletion See *mutation*.

denaturation The loss of the three-dimensional shape of a *protein* molecule. Denaturation changes the shape of the *active site* of an *enzyme*, reducing its effectiveness as a *catalyst*. Denaturation may be caused in several ways, including *pH* changes, heat, heavy metals and organic solvents, all of which break the weak bonds maintaining the secondary and tertiary structure of the protein. Mild denaturation by small pH changes may be reversible, but heat or treatment with heavy metals usually denatures proteins irreversibly, and leads to their coagulation. Enzymes stored for a long time also become gradually denatured.

dendrite A cytoplasmic extension of a nerve cell that carries a nerve impulse towards a cell body. Compare *axon*.

denitrification The process by which nitrogenous compounds and ions are broken down and gaseous nitrogen is released to the air. Dentrification takes place largely under anaerobic conditions by bacteria (denitrifying bacteria, e.g. *Bacterium denitrificans*) in soil and water and reduces soil fertility. See also *nitrogen cycle*.

denitrifying bacteria See *denitrification*.

density-dependence The regulation of the size of a population by factors whose effects increase in severity as population density increases. Such density-dependent factors are always biotic and include competition for limited resources (e.g. food and space). Compare *density-independence*.

density-independence The regulation of the size of a population by factors whose effects do not change in severity with changes in the density of the population, so they tend to affect the same proportion of the population irrespective of population density. For example, low temperatures during a severe winter may kill a fixed percentage of birds irrespective of the population density. Density-independent factors may be *abiotic* or *biotic*. Compare *density-dependence*.

dental formula A form of notation for describing the dentition of mammals. The formula consists of a series of fractions giving, in turn, the number of incisors (i), canines (c), pre-molars(p), and molars(m) on one side of the skull. The upper numbers represent the teeth in the upper jaw, and the lower numbers represent the teeth in the lower jaw. In humans, the dental formula is

$$2/2i \quad 1/1c \quad 2/2p \quad 3/3m.$$

dentine See *tooth*.

deoxyribonucleic acid See *DNA*.

deoxyribose A pentose sugar present in *DNA*. It is similar to ribose sugar but has one less atom of oxygen; an H atom rather than an OH group is attached to the second carbon atom.

depolarization A transient reversal in potential difference across the cell-surface membrane of a *neurone* (e.g. when an *action potential* is produced).

desertification The process by which deserts expand into adjacent, semi-arid lands. Climate change is the main cause of desertification, but the process may be accelerated or exacerbated by human activities in areas bordering deserts (e.g. overgrazing by domestic animals or removal of trees for firewood).

detergent An agent that acts on the surface of a substance (and is therefore called a *surfactant*) to remove or disperse grease and dirt. Detergents have a water-liking part and a water-hating part. When detergents are added to mixtures of oil and water, the oil is broken up into tiny droplets suspended in the water (an emulsion), with the water-liking part of the detergent in the water and the water-hating part in the oil. Detergents are often used to disperse oil that has been spilled accidentally. By breaking the oil into small droplets, dispersants increase the total surface area of the oil, thus accelerating the rate at which it is degraded (mainly by bacteria) into harmless compounds. This minimizes the risk of oil contaminating ground water, but detergents must be used with care. Some detergents and oil-detergent mixtures are poisonous to certain animals. See also *bile*.

detoxification The process by which organisms break down toxic substances chemically, or otherwise render them harmless. For example, hydrogen peroxide (a toxic by-product of *aerobic respiration*) is detoxified into water and oxygen by catalase, which is found in most cells. In mammals, detoxification of *amino acids* and many other substances takes place in the *liver*.

detritivore See *detritus*.

detritus Small pieces of dead and decomposing organisms, mostly plants. Organisms which feed on detritus (*detritivores*) may also consume the bacteria and fungi performing the decomposition, as well as other small organisms which use the detritrus as a habitat.

development The continuous changes in size, shape, form and degree of complexity that accompany the growth of a multi-cellular organism from a single cell to a mature adult.

diabetes See *diabetes mellitus*.

diabetes insipidus See *antidiuretic hormone*.

diabetes mellitus A metabolic disorder that results in excess sugar appearing in the blood, eventually resulting in coma and death if not controlled. It can be due to a variety of causes, including a decreased sensitivity to *insulin*, an inability to produce insulin, or overproduction of *hormones* that oppose the action of insulin. It is often referred to simply as diabetes.

dialysis The process by which small molecules (e.g. *salts* and *urea*) can be separated from large ones (e.g. *proteins* and polysaccharides) using a *selectively permeable membrane*. Small molecules pass through the membrane while large ones are prevented by their size from doing so. All *cell membranes* are selectively permeable and capable of dialysis. Dialysis is the basis of the process that separates urea from blood in the *kidney*.

diaphragm A sheet of muscular tissue, present only in mammals, which separates the thoracic cavity from the *abdomen*. It plays an important role in breathing. When relaxed, it is dome-shaped and projects into the *thorax*. During inhalation it contracts and flattens downwards to increase the volume of the thoracic cavity. During exhalation, it relaxes, regains its dome shape because of its own inherent elasticity, and reduces the volume of the thoracic cavity.

diastema A naturally occurring gap in the teeth such as that between the incisors and premolars of some herbivores such as hamsters which lack *canine teeth*.

diastole See *cardiac cycle*.

diastolic pressure See *blood pressure*.

dichotomous key A method used to identify and name an organism. The key consists of a series of paired statements (or questions) each referring to a pair of mutually exclusive, observable characteristics. The user must choose the statement appropriate to the organism being identified and in doing so is directed to another pair of statements and so on until the organism is identified.

dicots See *Dicotyledonae*.

Dicotyledonae, dicotyledons or **dicots** A class of *Angiospermatophyta* in which *embryos* have two *cotyledons* and a net-like (*reticulate*) pattern of veins on the leaves. Dicots include most trees, shrubs and herbs. Compare *Monocotyledonae*.

dicotyledons See *Dicotyledonae*.

diet The food an organism eats. A balanced diet is one comprising optimal amounts of food constituents in the correct proportions to promote good health. Although the precise quantities depend on the age, sex, size and activity of each individual, a balanced diet in humans should include *carbohydrates*, fats, *proteins*, minerals, *vitamins*, *roughage* and water.

differential centrifugation See *subcellular fractionation*.

differentially permeable membrane See *partially permeable membrane*.

differentiation The process, in a multicellular organism, by which unspecialized structures or cells become modified and specialized to carry out specific functions. Differentiation involves changes in both the structure and physiology of cells. In cells that become very specialized, the changes may be irreversible (e.g. the loss of a nucleus in red blood cells). Such highly specialized cells may be unable to carry out some basic functions and thus become dependent on other cells.

diffusion The net movement of molecules or ions from a region of high concentration to a region of low concentration. This movement continues as long as there is a *concentration gradient*. The rate of diffusion is affected by:

- the concentration gradient
- the area across which diffusion occurs
- the distance over which the diffusion takes place
- the structure across which diffusion occurs
- the size and type of diffusing molecule.

It is a passive process that uses the kinetic energy of molecules. See also *facilitated diffusion*.

digestion The breakdown of large complex organic compounds into smaller, simpler materials that can be absorbed and used in the *metabolism* of an organism. Digestion may be mechanical (e.g. the physical breakdown of plant cells by the grinding action of teeth) or chemical. Chemical digestion involves *enzymes* that catalyse *hydrolysis* reactions; the enzymes may be secreted on to the food, either in a gut or outside the body (*extracellular digestion*), or contained within *lysosomes* which digest food inside a cell (*intracellular digestion*).

dihybrid A doubly *heterozygote* organism (i.e. one that carries two different alleles for one *gene* and two different alleles for another gene). Mendel based his *law of independent assortment* on dihybrid crosses (matings between two dihybrids) which produced offspring having four different *phenotypes* in the ratio of 9:3:3:1. Dihybrid crosses involving linked genes, or genes which act to control the same characteristic, do not produce this simple Mendelian ratio.

dihybrid inheritance Inheritance involving two distinct *genes*. In simple *Mendelian inheritance*, the two genes affect two contrasting characteristics and are situated on different loci. Non-Mendelian inheritance may involve two linked genes situated on the same *chromosome* and the genes may act together to control the same characteristic.

dipeptide The product of joining together two *amino acids*. A *condensation* reaction takes place between the amino (NH_2) group of one amino acid and the carboxyl group (—COOH) of another, so that water is removed and a *peptide bond* is formed linking the two amino acids together.

diploblastic (of animals) Having two layers of cells (the *ectoderm* and *endoderm*) in the *embryo*, from which all other cells are derived. See *Cnidaria*.

diploidy The state of having body cells each of which contains two sets of *chromosomes* (i.e. the *diploid state*). The most persistent stage in the life cycle of most animals and plants is diploid. This allows a greater amount of genetic variability than would be possible in individuals with only one set of chromosomes. *Recessive* alleles that may be harmful in the present environment can persist in a population because, in *heterozygotes*, they are hidden from *natural selection* by their *dominant* counterparts. Recessive *alleles* are exposed to selection only when two of them occur in the same individual.

dipolarity The condition of molecules (e.g. water) that have an uneven distribution of electrical charge, so they contain areas with both very slight negative and very slight positive charges. In the case of water, the area around the hydrogen atom has a very slight positive charge and the area around the oxygen atoms has a slight negative charge. See also *hydrogen bonding*.

directional selection See natural selection.

disaccharide or **double sugar** A sugar formed when two *monosaccharides* are joined together by a reaction in which water is removed (*condensation*). For example, when *glucose* and *fructose* combine they form the disaccharide *sucrose*; glucose and galactose combine to form *lactose*; and two glucose molecules combine to form *maltose*.

disc See *vertebra*.

discontinuous variation See *variation*.

disease Any disorder or illness (but not injury) of the body or organ. In humans, each disease has a characteristic set of signs (features observable by a doctor) and symptoms (features perceived by the patient).

disruptive selection See *natural selection*.

distal (of a structure) Furthest away from the midline of an organism, or at the point furthest from the attachment to the main body (e.g. toes are at the distal end of the legs).

distribution 1. The geographical range of a species or population.
2. The spatial arrangement of organisms within an area. Individuals may be distributed randomly, regularly spaced apart, or grouped together.

diuretic See *caffeine*.

diversity See *species diversity*.

division See *taxon*.

division of labour See *multicellularity*.

DNA or **deoxyribonucleic acid** A complex *nucleic acid* which is responsible for inheritance in most living organisms. According to the model proposed by James Watson and Francis Crick in 1953, the DNA molecule is composed of two chains of nucleotides wound around each other to form a double helix. Each nucleotide consists of a base (adenine, cytosine, guanine or thymine) linked to a pentose sugar (deoxyribose) and a phosphate molecule. DNA is able to copy itself. The *hydrogen bonds* between the two polynucleotide chains are broken to expose the bases which act as templates for the formation of new polynucleotide chains. Thus, a new double-stranded DNA molecule is formed alongside an existing single strand by complementary base pairing. DNA replication requires energy in the form of *ATP* and special *enzymes* (called DNA polymerases) that initiate, control, and stop the replication. In *eukaryotes*, DNA replication occurs during *interphase* of the *cell cycle*. DNA carries hereditary information in its sequence of nucleotide bases (see *genetic code*); this determines the structure of proteins and hence indirectly controls all enzyme-driven reactions. DNA is found in *chromosomes*, *mitochondria*, and *chloroplasts* of eukaryotes, and in chromosomes and *plasmids* of *prokaryotes*.

DNA polymerase See *DNA*.

DNA replication See *DNA*.

dominance 1. (in genetics) The property possessed by *alleles* that are expressed in the *phenotype* of a heterozygous or homozygous individual. In complete dominance, one allele (the *dominant allele*) in a heterozygous individual completely masks the effects of the other allele; in codominance, both alleles are expressed to some degree.
2. (in ecology) The extent to which a particular species (the *dominant species*) predominates in a community at a particular time, and influences other species. Plants which are *dominant* often have a great influence on the characteristics of the environment and affect the composition of the community.

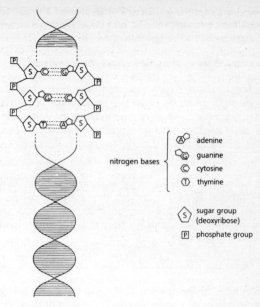

nitrogen bases
- adenine
- guanine
- cytosine
- thymine

sugar group (deoxyribose)

phosphate group

DNA

dominant See *dominance*.

dormancy An inactive state in which an organism has a low metabolic rate and growth is very slow or ceases. Dormancy may be imposed by unfavourable external conditions or be innate and occur as a normal part of the life cycle. Periods of innate dormancy usually coincide with an unfavourable season and, in plants, involve seeds and perennating organs (see *perennation*).

dormin See *plant growth substance*.

dorsal (of part of an organism) Furthest from the ground; for example, the upper surface of a leaf or wing, or, in vertebrates, the surface closest to the spinal cord.

double circulatory system See *circulatory system*.

double fertilization A process unique to flowering plants in which one male nucleus from the *pollen tube* fuses with the two polar nuclei of the *embryosac* to form the *endosperm*, and another male nucleus (the generative nucleus) fuses with the egg cell (ovum) to form the zygote which develops into the embryo.

double helix The three-dimensional structure of *DNA* (and double-stranded *RNA*), in which two *polynucleotide* chains intertwine to form a right-handed helix. The chains are held together by *hydrogen bonding* between two complementary bases (see *base pairing*).

double sugar See *disaccharide*.

Down's syndrome Symptoms that occur in human beings who carry three copies of chromosome 21 instead of two. The syndrome includes moderate to severe mental retardation, a small, round head, slanting eyes, and minor abnormalities of the feet and hands. It is caused by incorrect separation of *homologous chromosomes* during *meiosis* and the formation of *gametes*, or incorrect **cell division** in the early development of the *embryo*. Down's syndrome can be detected before birth by means of *amniocentesis*.

drone A male social insect, especially a bee, whose only function is to mate with fertile females.

Drosophila A genus of fruit flies that includes the species *Drosophila melanogaster*, used extensively for genetics research.

drug Any substance that alters the natural internal chemical environment of an organism and affects its normal body functions.

duct A small tube, composed of cells, that conveys fluid or other material.

duodenum The first part of the *small intestine*, between the *stomach* and *ileum*, into which partially digested substances (*chyme*) pass from the stomach. Digestion continues in the duodenum with the aid of juices secreted by the *pancreas*, *bile* from the *liver*, and other intestinal juices secreted from the highly folded walls of the duodenum. *Villi* (microscopic, finger-like projections) in the duodenal wall increase the surface area for digestion and absorption. Small, rounded glands (*Brunner's glands*) in the *submucosa* (a deep layer) of the duodenum secrete alkaline salts and mucus which probably help protect the duodenal wall from acids and proteases in the chyme and also provide an alkaline environment necessary for the efficient action of duodenal *enzymes*.

duplication See *mutation*.

dynamic equilibrium The condition of a system that is stable because factors working in opposing directions balance each other. In a system undergoing a reversible chemical reaction, it is the state reached when the reaction rates in the two opposing directions are equal, so there is no overall change.

ear The organ of hearing and balance in vertebrates. In mammals, the ear comprises the *pinna*; the outer ear or auditory canal; and the middle ear and inner ear which are contained within the skull. The pinna consists of a flap of tissue which helps direct sound into the outer ear (auditory tube), leading to the *eardrum* (or tympanic membrane). The eardrum is a thin, double membrane of *epidermis*; sound waves vibrate it and these vibrations are transmitted through the middle ear by the *ear ossicles* to the oval window (*fenestra ovalis*), separating the middle ear from the inner ear. The three ear ossicles act as levers; if a very loud sound enters the ear, muscles attached to the ear ossicles prevent them vibrating too much. Vibrations of the last ossicle (the *stapes*) against the oval window causes waves to pass along the fluid in the *cochlea;* this contains receptor cells that *transduce* (change) the sound waves into nerve impulses which pass along the auditory nerve to the brain, different parts of the cochlea responding to sounds of different pitch (height or depth of sound, which depends on the frequency of the sound waves). In addition to the cochlea, *semicircular canals* also occur in the inner ear and are involved in maintaining balance and body posture.

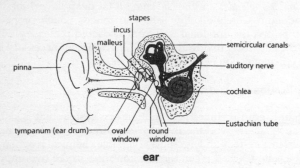

ear

eardrum See *tympanic membrane*.

ear ossicle One of usually three small bones (*malleus, incus* and *stapes*) that transmit vibrations in the middle *ear* in mammals.

Echinodermata or **echinoderms** A phylum of spiny-skinned, invertebrate animals that are entirely marine. Their bodies usually have a five-way symmetry and they use a water-vascular system with tube feet for locomotion.

echinoderms See *Echinodermata*.

E. coli See *Escherichia coli*.

ecological efficiency The efficiency with which energy is transferred from one *trophic level* to the next. It is the percentage of energy in the *biomass* produced by one trophic level that is incorporated into the biomass produced by the next higher trophic level.

ecology The scientific study of organisms in relation to all aspects, living and non-living, of their natural environment. Autecology focuses on relationships between individual organisms or populations and their environment; synecology deals with communities and the environment.

ecosystem A discrete, relatively stable and self-contained system comprising a community of organisms and their abiotic and biotic environment (e.g. a pond, rocky shore, oak woodland). Members of the community are linked by feeding relationships and are, in varying degrees, interdependent.

ectoderm The layer of cells on the outer surface of an *embryo*. The skin of animals is derived from the ectoderm.

ectoparasite See *parasite*.

ectotherm An animal whose source of body heat is largely external. Some ectotherms can regulate their body temperature within a fairly narrow range by behavioural means (e.g. by basking in the sun or hiding in the shade). Compare *endotherm* and *poikilotherm*.

edaphic factor A factor that affects the distribution of organisms and is related to the biological, physical or chemical composition of the soil. Examples are mineral content, pH, water content, humus content).

effector A structure (cell, organ, etc.) that responds to a hormonal stimulus or stimulus from a nerve impulse by carrying out a particular function (e.g. a muscle or gland).

egestion The elimination from the body of undigested food that has never left the gut. Unlike excreted material, egested material has not been metabolized.

egg, egg cell or **ovum** The female *gamete* that contains a single set of *chromosomes* (i.e. it is haploid). In flowering plants, it is contained within the *embryosac*. In animals, it is often packed with nutritive *yolk* granules.

egg cell See *egg*.

ejaculation The discharge of a secretion with some force. It especially refers to the forcible emission of seminal fluid through the penis of an animal.

electromagnetic spectrum See *light*.

electron A negatively charged particle with very small mass. Electrons form the outer part of an atom.

electron acceptor A molecule that readily accepts an electron. In photosynthesis, for example, an electron acceptor captures the electron lost from *chlorophyll* when it is excited by light energy, and oxygen is the final electron acceptor in *aerobic respiration*. Many electron acceptors are also electron carriers and readily pass on the electron to another molecule. The molecule gaining the electron becomes reduced, the one losing the electron becomes oxidized. These reduction—oxidation (redox) reactions are important in photosynthesis and respiration. See also *electron transport system*.

electron carrier See *electron acceptor*.

electron micrograph See *microscope*.

electron microscope See *microscope*.

electron transport chain See *electron transport system*.

electron transport system or **electron transport chain** A series of redox reactions in each of which an electron or hydrogen atom passes from one molecule to another (electron carriers), releasing energy used to synthesize *ATP*. Electron transport systems are important in *aerobic respiration* and in the *light-dependent stage* of *photosynthesis*.

electro-osmosis See *translocation*.

electrophoresis A technique for analysing and separating particles of different electrical charges. It is used extensively to study mixtures of macromolecules (e.g. *proteins* and *nucleic acids*). Typically, the apparatus consists of a supporting medium soaked in a suitable *buffer* with an electric field set up across it. The mixture to be separated is placed on the supporting medium. The components with a positive charge migrate towards the cathode, and those with a negative charge towards the anode. The rate of migration varies from one substance to another, depending on the size, shape, and charge of its molecules, thus the molecules separate from each other. Their final position is compared with the positions of known standards.

element A substance that cannot be split into simpler substances by chemical means. The most common elements in living organisms are carbon, hydrogen, oxygen and nitrogen.

Elodea canadensis or **Canadian pond weed** A freshwater flowering plant that has a rapid rate of *photosynthesis* in bright light and produces large amounts of oxygen.

embedding Supporting a biological specimen in a substance so the specimen can be cut into thin sections. Specimens used in light microscopy are usually embedded in paraffin wax, but those used in electron microscopy must be cut very thinly and resin is used as an embedding medium, because it provides much firmer support.

embryo A young multicellular organism, formed after the fusion of a male and female **gamete**, in the first stages of development. In flowering plants, the embryo is contained within a seed and consists of one or two cotyledons (seed leaves), a plumule (seed shoot), and a radicle (seed root). In animals, the embryo is the stage contained within the egg or reproductive structures of the mother between fertilization and hatching or birth. A human embryo is called a *foetus* after the first eight weeks of pregnancy, at which time the main adult features can be recognized.

embryosac The female *gametophyte* of flowering plants. It is a sac-like structure produced in the carpel and is the site of *double fertilization* and development of the *embryo*. It consists of eight haploid nuclei that include an egg cell (the female gamete) and two polar nuclei that form the *endosperm*.

emphysema A degenerative disease in humans, in which lung tissue loses its elasticity. Air tends to be trapped in the lungs, making breathing more difficult.

emulsification The formation of an emulsion (i.e. small droplets of oil or fat suspended in water). See also *bile*.

emulsion test A test for *lipids* in which the test substance is added to absolute ethanol. Distilled water is then added to the mixture after it has been left a sufficient time for fats to dissolve in the alcohol. When the mixture is shaken, it appears cloudy if fats are present, because of the formation of droplets of fat (emulsion) in the water. The test is usually performed on solids or on liquids suspected of being oils. It is not appropriate for watery substances.

enamel An extremely hard, smooth, white material consisting mainly of crystals of calcium phosphate and calcium carbonate salts bound together by *keratin* fibres. Enamel is formed from the *epithelium* and lines the exposed parts of teeth and the denticles of fish.

endangered species A species whose numbers have been reduced to a critical level or whose *habitat* has been so drastically reduced that it is deemed to be in imminent danger of extinction unless remedial action is taken, such as an artificial breeding programme or improvement of the habitat.

endergonic reaction A reaction that uses *free energy*.

endocarp See *fruit*.

endocrine gland A ductless gland that manufactures *hormones* and secretes them directly into the blood, which transports them to their destination. Endocrine glands tend to control slow, long-term activities (e.g. growth and sexual cycles). In mammals, the activity of many endocrine glands is controlled by negative feedback mechanisms coordinated by the *pituitary* gland.

endocuticle See *cuticle*.

endocytosis An active, energy-consuming process that results in the bulk transport into the *cytoplasm* of a cell of substances too large to pass through the cell surface membrane by *diffusion*, *facilitated diffusion*, or *active transport*. The *cell membrane* invaginates (folds inwards) to surround the substance and pinches off to form an intracellular vesicle or vacuole. Endocytosis of solid substances is called *phagocytosis* ('cell-eating'). Cells specializing in this process (e.g. *white blood cells*) are called phagocytes (or scavenger cells). Endocytosis of liquids is called *pinocytosis* ('cell-drinking').

endoderm The innermost layer of cells in an *embryo*. In *Cnidaria* it gives rise to the inner epithelium (see *epithelia*) lining the gut cavity. In *triploblastic* animals, it lines the inside of the stomach and intestines.

endoparasite See *parasite*.

endopeptidase An *enzyme* that accelerates the breakdown of proteins by catalysing the *hydrolysis* of internal *peptide bonds*.

endoplasmic reticulum A system of parallel membranes enclosing fluid-filled channels (*cisternae*, sing. *cisternum*) within the *cytoplasm* of animal and plant cells. A smooth endoplasmic reticulum has no *ribosomes* attached to the membranes, a rough endoplasmic reticulum does. The channels of the endoplasmic reticulum form a transport system within cells.

endoskeleton A *skeleton* contained entirely within the body of an animal. In vertebrates, the endoskeleton is made from bone or *cartilage*. It is usually jointed and muscles are attached to allow movement. In addition to providing a system of levers on which muscles can act, the endoskeleton also protects delicate organs and provides a supporting framework for soft structures.

endosperm The layer of tissue that surrounds the *embryo* in the seeds of many flowering plants, acting as a food source for the developing embryo. Endosperm cells are triploid (i.e. each cell has three sets of *chromosomes*). During *double fertilization*, a primary endosperm nucleus is formed by the fusion of two haploid nuclei of the *embryosac* with one male gamete. Subsequent division by *mitosis* of the primary endosperm nucleus produces the endosperm.

endotherm An animal that can generate body heat and regulate its core temperature by internal, physiological mechanisms despite variations in environmental temperature. Birds and mammals are endotherms. Compare *ectotherm*.

energy The capacity to do work, which is the product of force and distance (work = force x distance). Energy is measured in **joules**; 1 joule of work is done when a force of 1 newton moves through a distance of 1 metre. See also *free energy* and *thermodynamic laws*.

energy budget The balance of *energy* input and output in a biological system (an organism, *trophic level*, or *ecosystem*). An energy budget can be expressed in an equation:

$$C = P + R + G + U + F,$$

where C is consumption (the total intake of food or energy), P is growth (that part of energy input retained and incorporated into the biomass), R is respiration (that part of energy input lost as heat either directly or through the mechanical work performed), G is that part of energy input released as reproductive bodies, U is that part of energy input that is absorbed but later lost in excreted or secreted materials, and F is that part of energy input lost as faeces. All components of the energy budget should be expressed in joules.

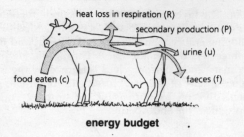

energy budget

energy flow The movement of *energy* through an *ecosystem*. Energy is usually first trapped by producers as radiant energy from the Sun; it is then transferred from one *trophic level* to another, some *free energy* being lost as heat from the ecosystem at each stage of the process. See also *pyramid of energy*.

enteron Any gut cavity, but especially the sac-like gut cavity of coelenterates (e.g. *Hydra*). See *Cnidaria*.

environment The surroundings of an organism; the sum total of the external conditions, biotic and abiotic, in which an organism lives. See also *internal environment*.

environmental resistance The combined effect of all the environmental factors that limit growth of populations. These depend on the species and circumstances, but include shortage of food, lack of light, predators, lack of living space and shelter, disease, accumulation of metabolic wastes, an unfavourable climate, and, in some species, behavioural changes due to the stress associated with overcrowding.

enzyme A *protein* molecule that acts as a biological *catalyst* for a reversible chemical reaction. (A catalyst is a substance that, in very small amounts, speeds up a chemical reaction without itself being used up or altered in the reaction.) Enzymes cannot make reactions occur which otherwise would not happen, and they do not alter the final amount of product formed. An enzyme combines with a specific *substrate* at its *active site* to form an enzyme—substrate complex (see *lock-and-key theory*). This complex follows an alternative pathway for the chemical reaction that has an *activation energy* lower than that for the pathway without the enzyme. Once the reaction is completed, the product detaches from the enzyme which, therefore, can be used over and over again.

epicarp See *fruit*.

epicuticle See *cuticle*.

epidermis 1. The outermost layer of animal cells, derived from the *ectoderm*. In invertebrates, it is one cell thick, in some animals (e.g. insects) being covered by an impermeable *cuticle*. In vertebrates, it is the outer layer of skin, several cells thick, which may contain a variety of specialized structures (e.g. scales, hairs, or feathers). In land-dwelling animals, its surface layer consists of dead, hardened cells, impermeable to water. **2.** A continuous, compact layer, one cell thick, on the outermost surface of a plant. In ferns and terrestrial seed-bearing plants, the epidermis of aerial parts contains *stomata* and is lined by a waxy cuticle. The epidermis forms a thin 'skin', providing some mechanical support and helping to protect underlying tissues from physical damage and water loss.

epiglottis In mammals, a moveable flap of *cartilage* on the ventral wall of the *larynx* that covers the opening to the *trachea* when food is swallowed.

epinephrine See *adrenaline*.

epithelium (*pl* **epithelia**) A sheet or tube, consisting of one or a few layers of closely packed cells, lining the external and internal surfaces of organs and the walls of body cavities in multicellular animals. It also forms glands and some parts of sense organs. The function of epithelia varies according to their structure and location, and may be protective, secretory, sensory or absorptive.

equator See *spindle apparatus*.

erosion The process by which the products of the weathering of rocks and soil are transported by the action of wind, running water, moving ice etc.

erythrocyte See *red blood cell*.

Escherichia coli or *E. coli* A rod-shaped bacterium used extensively in the study of genetics and molecular biology. It is a normal inhabitant of the human gut and is usually harmless, although some strains can cause disease. *E. coli* is used frequently in genetic engineering. After the application of *recombinant DNA* technology, it can be cultured in fermenters to produce non-bacterial substances (e.g. human insulin).

essential amino acid See *amino acid*.

ester A compound formed by a *condensation* reaction that links an acid and an alcohol. Fats and oils are esters of *glycerol* and *fatty acids*. See also *lipid*.

ester bond A —COO— bond that links an acid to an alcohol in an *ester*.

esterification See *lipid*.

ethanol fermentation See *fermentation*.

ethene See *plant growth substance*.

ethylene See *plant growth substance*.

etiolation The form of growth observed in plants which grow in the dark. Etiolated plants are typically tall because they have long *internodes*, spindly because they have poorly developed lignified tissue, and yellow because they lack *chlorophyll*.

euchromatin See *chromatin*.

eukaryote Any organism that contains a cell or cells with a well-defined *nuclear envelope* surrounding the nucleus. The cells also have double-membraned *organelles* (e.g. *mitochondria*) and the DNA is complexed with proteins to form *chromosomes*. All organisms except bacteria are eukaryotes. Compare *prokaryote*.

euploidy See *mutation*.

Eustachian tube A tube connecting the middle *ear*, in vertebrates, with the back of the throat. It is normally closed, but is opened by swallowing or yawning. This equalizes the air pressure to either side of the ear drum by allowing air to enter the middle ear. It is named after the Italian anatomist Bartolommeo Eustachio (*c.* 1520—74).

eutrophication The enrichment of water by the addition of nutrients. Eutrophication occurs naturally, but is also caused by pollution. The additional nutrients encourage the rapid growth of annual plants, especially algae, which may smother higher plants, reduce light intensity, and produce toxins which kill fish. When the plants die, bacterial decomposition may use up all the oxygen dissolved in the water, leading to the death of organisms dependent on oxygen for *aerobic respiration*.

evapotranspiration The loss of moisture from a leaf surface by means of direct evaporation (change of liquid water into water vapour) linked with *transpiration*. Evapotranspiration makes a significant contribution to the formation of rain clouds in tropical rain forests. See also *water cycle*.

evolution or **organic evolution** The development of new species of organisms from pre-existing ones, the process by which all present-day organisms are thought to have developed from common ancestors. It results from changes in the genetic composition of a population occurring over successive generations. *Microevolution* is the term used for changes in allele frequencies occurring within a species over a relatively short time (e.g. the acquisition of pesticide resistance by insects). *Macroevolution* is the term for evolutionary changes above the species level, resulting in the formation of new higher taxonomic groups (e.g. genus and class). Modern versions of Darwin's theory of evolution by natural selection (see *Darwinism*) are generally regarded as describing the most probable mechanism by which evolution occurs.

excretion The elimination from the body of metabolic waste products (e.g. carbon dioxide and nitrogenous substances such as *urea*). Plants have no regular excretion of nitrogenous wastes, because they synthesize only the *amino acids* they require. Animals depend on external sources of amino acids and inevitably absorb some they do not require and which must be excreted because they cannot be stored. In mammals, the main organs of nitrogenous excretion are the *kidneys*, but some nitrogenous wastes are also excreted in sweat. The lungs excrete carbon dioxide.

exergonic reaction A reaction that releases *free energy*.

exhalation or **expiration** The act of forcing air or water out of a respiratory organ. In mammals, for example, air is exhaled through the mouth and nose. See also *ventilation*.

exine See *pollen grain*.

exocrine gland A gland that secretes a substance into a duct carrying it to its destination. The *pancreas* contains exocrine glands that produce pancreatic juices; pancreatic ducts carry the juices to the *duodenum*, where they help digest food.

exocuticle See *cuticle*.

exocytosis The emptying of the contents of a membrane-lined vesicle or vacuole at the surface of a cell by fusion with the cell surface membrane.

exopeptidase An *enzyme* that accelerates the breakdown of proteins by catalysing the *hydrolysis* of *peptide bonds* at the end of polypeptide chains, thereby successively splitting off the terminal *amino acids*.

exophthalmic goitre See *thyroid gland*.

exoskeleton A *skeleton* that lies outside living tissue. The exoskeleton of *arthropods* is a rigid, external covering of hard chitinous material (see *cuticle* (2)). An exoskeleton protects and supports the body. Projections on the inside of the skeleton (called *apodemes*) provide points of attachment for muscles. In order to grow, the exoskeleton of arthropods must be shed at intervals. Most molluscs have an external skeleton consisting of one or two shells, but these do not enclose the animal completely and therefore need not be shed for growth to take place.

exotherm See *poikilotherm*.

experiment An investigation designed to examine a *hypothesis* by testing predictions based on it.

expiration See *exhalation*.

external environment The environment outside the body of an organism.

exteroceptor See *receptor cell*.

extinction The complete elimination of a species or larger taxonomic group. Some scientists believe mass extinctions are cyclical events related to meteor showers, but this idea is not universally accepted and rival theories are based on sea-level changes and climate change, although the mass extinction which ended the Cretaceous Period (about 65 million years ago) is widely believed to have been caused by the impact of a large asteroid.

extracellular digestion See *digestion*.

eye The organ of sight, which is sensitive to the direction and intensity of light. In mammals, eyes are nearly spherical structures occurring in pairs. A tough, fibrous, non-elastic layer (sclerotic layer), continuous with the *cornea*, surrounds and protects the eye. The *choroid layer*, a pigmented layer between the *retina* and sclerotic coat, reduces internal reflections within the eye; it contains *blood vessels* supplying the eye with oxygen and nutrients. At the front of the eye, the choroid layer is modified to form the *ciliary body* and *iris* (a ring of opaque-tissue containing circular and radial muscles which can contract or relax to vary the size of the pupil). A watery fluid (*aqueous humour*) secreted by glands in the ciliary body fills the space between the cornea and *lens* and a jelly-like fluid (*vitreous humour*) fills the space behind the lens; these fluids help maintain the shape of the eye. Light entering the eye is refracted by the curved, transparent cornea before passing through the pupil, an aperture in the centre of the iris. The iris tends to be darkly pigmented in diurnal animals (e.g. dogs) and pale in nocturnal ones (e.g. cats). The light next passes through the lens, a transparent, biconvex, crystalline disc containing dead, prism-shaped cells filled with a jelly-like protein (*crystallin*), and held in position by *suspensory ligaments* (strong fibrous tissue). The shape of the lens can be altered by *ciliary muscles* to focus light on to the retina. In land animals, most of the focusing power is provided by the cornea, the lens being concerned only with fine-focusing of the image. The retina contains light-sensitive cells (*rods* and *cones*); the cones are most concentrated in the *fovea*. Rods and cones transduce (change) light energy into nerve impulses which are transmitted along the optic nerve to the visual area on the brain cortex where an image is perceived. A blind spot occurs where the optic nerve leaves the eye. See also *vision*.

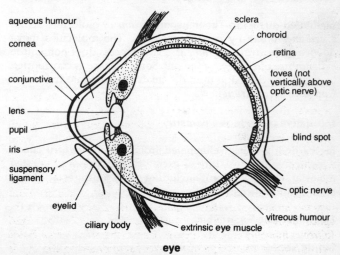

eye

F_1 A symbol commonly used in displays of solutions to genetics problems. It should be used only to designate the off-spring of homozygous parents. First-generation offspring of other types of crosses should be described more fully (e.g. as offspring (1) *phenotypes*, or offspring (1) *genotypes*).

F_2 A symbol commonly used in displays of solutions to genetics problems. It should be used only to designate the off-spring produced by crossing the F_1 parents. Second-generation offspring of other crosses should be described more fully (e.g. as offspring (2) *phenotypes*, or offspring (2) *genotypes*).

fabil An iodine-containing stain used on plant sections to aid in the examination of vascular and strengthening tissues. The staining reaction varies, but *xylem* vessels generally stain brown, *sclerenchyma* pink, *cellulose* pale blue, *cytoplasm* and nuclei dark blue, and *starch* black.

facilitated diffusion The transport of *ions* or molecules across a *cell membrane* by *carrier molecules*. The molecules move down a *concentration gradient*, so the process does not require expenditure of energy by the cell. Some *glucose* is transported by facilitated diffusion from the small intestine into the blood. Compare *active transport*.

facultative parasite See *parasite*.

faeces Bodily waste, eliminated from the *alimentary canal*, which contains a mixture of excretory products and egested material (see *egestion*), including *bile*, undigested food, bacteria and mucus.

Fallopian tube See *oviduct*.

family See *taxon*.

fatigue See *accommodation*.

fatty acid A component of most natural *fats* and *oils*. It consists of a long, linear chain of hydrocarbons with a carboxyl group (—COOH) at one end. Saturated fatty acids have the general formula $R—(CH_2)_n—COOH$, where R represents a hydrocarbon (e.g. —CH_3 or C_2H_5). Fatty acids which are loosely attached to blood *proteins* in our bodies are called *free fatty acids*; they are an important source of energy for long-duration activities. In saturated fatty acids, all the four potential connecting points on the carbon are occupied with four single covalent bonds. In unsaturated fatty acids, one or more double covalent bonds occur in the hydrocarbon chain. This produces a kink in the hydrocarbon chain making the molecule asymmetrical and less easy to pack tightly together than a molecule of a saturated fatty acid.

fauna All the animals found within a particular geographical region. Compare *flora*.

feather One of the structures found in birds, that provides the body covering. It is composed of the protein *keratin* and formed as an outgrowth of the *epidermis*. Feathers provide insulation, repel water, help the bird fly, and in some species are used in displays.

feedback Control of the performance of a system by part of the output of that system. Feedback may be negative or positive. In negative feedback, a system which deviates from a relatively constant state is brought back to that state. Many metabolic pathways are controlled by a form of negative feedback (sometimes called feedback inhibition) in which a product of the pathway inhibits one or more of the enzymes involved in the pathway. In positive feedback, control operates in the other direction, so small deviations in output are amplified. See also *homeostasis*.

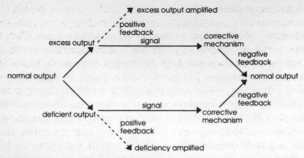

feedback Generalized diagram for any homeostatic control process using negative feedback.

feedback inhibition See *feedback*.

femur The thigh bone of terrestrial vertebrates, joined at one end to the *pelvic girdle* and at the other end to the *tibia* (shin bone). It is the largest and strongest bone in the body.

fenestra ovalis See *oval window*.

fenestra rotunda See *round window*.

fermentation 1. A synonym of *anaerobic respiration* and *glycolysis*. In animals, anaerobic respiration results in the production of *lactic acid* and is sometimes called lactic acid fermentation. In plants and fungi, anaerobic respiration results in the production of ethanol and carbon dioxide and is sometimes called ethanol (or alcohol) fermentation. Ethanol fermentation of sugars in yeast is exploited in the brewing and wine-making industries and in bread-making (the carbon dioxide making the dough rise).
2. Any process that involves culturing cells or microorganisms

in a *fermenter* (also called *bioreactor*), a chamber in which the environmental conditions (temperature, nutrients, oxygen concentration, pH, wastes, etc.) can be precisely controlled to achieve the maximum yield of the cell or microorganism, or one or more of their products. In batch processing, nutrients and microorganisms are placed in a closed fermenter and nothing is added to or taken from the fermenter during fermentation. Environmental conditions, other than temperature, are not ordinarily controlled. At the end of fermentation, the product is separated from the microorganisms. In continuous flow processing, nutrients are added continuously to the fermenter at a rate which equals their removal by micro-organisms. The microorganisms are maintained in optimum environmental conditions (including optimal population density) to provide maximum yield of microorganisms and/or their products, which are continuously removed from the fermenter.

ferns See *Filicinophyta*.

fertilization 1. The process in sexual reproduction in which the nuclei of two haploid *gametes* fuse to form a diploid *zygote*. Cross-fertilization occurs when the gametes are from different organisms; self-fertilization occurs when both gametes are from the same individual. Internal fertilization occurs if the fusion of gametes takes place inside an organism (as in humans); external fertilization if they fuse outside the organism (as in most bony fish and amphibians). See also *double-fertilization*.
2. The application of substances (fertilizers) in **agriculture** and horticulture, which supply plant nutrients and so stimulate crop growth.

fertilizer Any substance that is applied to soils as a source of nutrients for plant growth. Natural fertilizers (sometimes called organic fertilizers) include farmyard manure, crop residues, and *compost*. Most artificial (i.e. industrially manufactured) fertilizers contain salts of nitrogen, phosphorus and potassium, three mineral elements that are commonly deficient in farm soils. Commercial fertilizers are often marked with a three-number code that indicates the proportions of these three minerals (e.g. a fertilizer marked '10—12—8' is 10% nitrogen (as ammonium or nitrate), 12% phosphorus (as phosphoric acid), and 8% potassium (as potash). Excessive use of fertilizers may result in soluble salts (e.g. nitrates) leaching into water courses and causing *eutrophication*.

fibre 1. An elongated plant cell. See *sclerenchyma*.
2. Any thread-like structure (e.g. muscle fibre, nerve fibre).
3. An indigestible component of food, made mainly from *cellulose*. See also *roughage*.

fibrin An insoluble fibrous *protein* formed during *blood clotting* by the action of *thrombin* on *fibrinogen*. Fibres of fibrin form a mesh that traps red blood cells in the clot.

fibrinogen A large, soluble *protein* found in blood plasma. It is formed in the liver and converted to insoluble *fibrin* during *blood clotting*.

fibrous protein A *protein* consisting of chains of polypeptides which are mainly linear and strengthened by cross-linkages (e.g. *keratin*, elastin and *collagen*). They are relatively insoluble in water and very strong.

fibula The thin, outer bone in the lower leg of terrestrial vertebrates, between the knee and ankle.

Fick's law A law stating that the rate of *diffusion* in a given direction across an exchange surface is directly proportional to the area of the exchange surface and to the concentration difference across this surface, and inversely proportional to the length of the diffusion pathway.

fight or flight response See *adrenaline*.

filament 1. (in animals) Any thin, thread-like structure.
2. (in a flower) The stalk of a stamen to which an anther is attached.
3. A chain of cells (e.g. in *Spirogyra*, which is consequently called a filamentous alga).

Filicinophyta, Filicopsida or **ferns** A subphylum (or subdivision) of vascular plants (*tracheophytes*) in which the young leaves are coiled in a bud called a 'fiddlehead'. The structures in which asexual spores are formed (*sporangia*) occur in clusters called *sori* (sing. *sorus*).

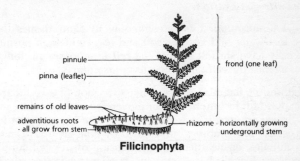

Filicinophyta

Filicopsida See *Filicinophyta*.

fimbriae See *pili*.

fish See *Chondrichthyes, Osteichthyes*.

fitness A measure of the ability of an organism to survive and produce offspring which will themselves be able to reproduce. The 'fittest' individual within a population is the one which produces the largest number of offspring that survive to reproduce themselves.

five-kingdom classification See *kingdom*.

fixation 1. (in the preparation of biological specimens for microscopical examination) The process of stabilizing (or fixing) the specimen in as natural a state as possible before *embedding* and sectioning. A chemical (fixative) is usually used (e.g. glutaraldehyde for light microscopy and osmium tetroxide for electron microscopy).
2. (in genetics) The possession by all members of a population of a particular *allele* of a *gene*, so no alternatives exist at that *locus* and the population is homozygous for that allele.
3. See *nitrogen fixation*.

flaccid Limp; applied most commonly to plant tissues that have lost water and become soft and less rigid than normal. The cells have shrunk so that the *cell membranes* no longer press against the *cell walls*. See also *plasmolysis*.

flagellum (*pl.* **flagella**) A thread-like extension of a cell that is specialized for locomotion or for movement of fluids close to a cell. In *eukaryotes*, it is formed from a core of *microtubules*: two central microtubules surrounded by nine groups of outer microtubules, ensheathed by an extension of the cell surface membrane. Flagella have a structure similar to cilia (see *cilium*) but are generally longer and occur in smaller numbers on each cell. *Bacteria* also have flagella, but they are different in both structure and mode of operation from the flagella of eukaryotes. Bacterial flagella rotate; eukaryotic flagella undulate.

flatworms See *Platyhelminthes*.

flora All the plants found within a particular geographical region.

florigen See *plant growth substances*.

flower The structure in *Angiospermatophyta* (flowering plants) that contains the sexual reproductive organs. Flowers are adapted to bring about pollination and fertilization, leading to the production of *seeds* and *fruits*. Typically, a flower consists of a receptacle, calyx, corolla, androecium and gynaecium. The *receptacle* is the swollen flower stalk to which other parts of the flower are attached. The *calyx* is the outermost part of the flower, containing the sepals. These are usually green, often hairy, leaf-like structures. The *corolla* consists of the petals. They are often spectacular and brightly coloured and scented in insect-pollinated plants. There is often a glandular swelling at the base of the petal (a *nectary*), that produces a sweet sugary solution (nectar) to attract insects. In wind-pollinated flowers, petals are usually a drab green colour and reduced in size. The *androecium* consists of all the male parts: the *stamens*, each of which comprises a stalk-like filament and an *anther*. The anther produces pollen, contained within four pollen sacs. The gynaecium comprises all the *carpels*, the female reproductive structures. Each carpel comprises a *stigma* attached via a stalk-like style to an *ovary*, which contains the *ovules*. If there is more than one carpel in a flower, they may be fused or separate. The stigma is the sticky female part to which pollen becomes attached during pollination. In wind-pollinated plants, it is usually pendulous (hanging downwards) and feathery. The ovules contain the female *gametes* within an *embryosac*. After fertilization, the ovule develops into the seed, and the wall of the ovary becomes the fruit.

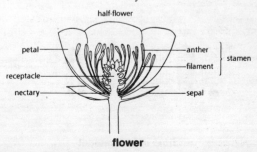

flower

flowering plants See *Angiospermatophyta*.

fluid-mosaic model A model suggesting that *cell membranes* consist of a mosaic of proteins floating in a fluid *phospholipid* layer. The phospholipids form a *bilayer* (i.e. a layer two molecules thick) because of their hydrophilic ('water-loving') phosphate heads and hydrophobic ('water-hating') *fatty acid* tails. Some proteins are confined to the inner or outer surface of the membrane, while others penetrate all the way through. Thousands of different proteins can occur in cell membranes and they have a variety of functions. Some are purely structural, strengthening the membrane, some are involved in transporting materials through the membrane, acting as carrier molecules or providing channels or pores through which water-soluble substances can pass, and others may act as *enzymes*, specific receptor molecules, *electron carriers* and energy transducers (in, for example, *respiration* and *photosynthesis*). Glycoproteins (proteins joined to a carbohydrate group) occur on the outside of the cell surface membrane to form a layer called the *glycocalyx*; the carbohydrate portions project from the membrane like antennae and play an important role in cell recognition and interactions between cells (for example, glycoproteins act as *antigens* in the immune response).

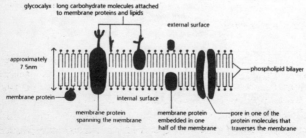

glycocalyx : long carbohydrate molecules attached to membrane proteins and lipids

external surface

approximately 7.5nm

phospholipid bilayer

membrane protein

internal surface

membrane protein spanning the membrane

membrane protein embedded in one half of the membrane

pore in one of the protein molecules that traverses the membrane

fluid mosaic model

fluoride A compound of fluorine that is often naturally present in drinking water. Small amounts of sodium fluoride may be added to drinking water lacking natural fluoride to prevent tooth decay. The fluoride becomes incorporated in tooth *enamel* making the teeth more resistant to dental caries.

focusing or **accommodation** The adjustments that occur in the *eye* in mammals, to allow distant and near objects to be seen clearly, achieved mainly by changes in the shape of the lens, assisted by changes in the size of the pupil.

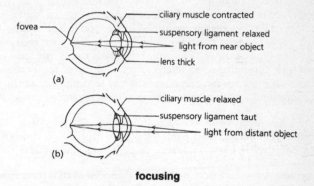

focusing

foetus See *embryo*.

folic acid A member of the water-soluble vitamin B complex (see *vitamin B1 and vitamin B2*). Folic acid acts as a coenzyme (see **cofactor**) in the synthesis of some *amino acids* and *DNA* and is important for the formation of normal red blood cells. Deficiency causes reduced growth and anaemia. Folic acid is manufactured by symbiotic gut bacteria. Rich dietary sources of folic acid are *yeast* and green vegatables.

follicle stimulating hormone A *hormone* secreted by the anterior *pituitary* gland in mammals, which stimulates the growth of the *Graafian follicles* of the ovary of a female, and the production of sperm in the testes of males. It is a major ingredient of fertility drugs used to increase ovulation and sperm production.

food chain A representation of the transfer of energy through a series of organisms in a community. A food chain shows a sequence of organisms, usually starting with a producer and ending with the top consumer, in which each organism is the food of the next member in the chain. The number of links in a food chain is usually limited to four or five, because a high percentage of the *free energy* consumed in the food is lost as heat at each stage. The food chain assumes that each organism feeds on only one type of other organism, but this is rarely the case. A *food web*, consisting of an interconnected group of food chains, is a much more realistic representation of feeding relationships of a community within an *ecosystem* and *energy flow* through it.

food poisoning An illness affecting the digestive system that results from eating food contaminated with harmful bacteria (e.g. *Salmonella*), bacterial toxins, or poisonous chemicals.

food test A method that identifies a component of a food mixture. Common food tests include *Benedict's test* for reducing sugars, *Biuret test* for proteins, *emulsion test* for fats, 'iodine' test (see *starch test*) for starch, *phloroglucinol* test for lignin, and Schultze's test (see *Schultze's solution*) for cellulose.

food web See *food chain*.

fossil The remains or traces of an organism that lived in the past and which has been preserved by natural processes (in, for example, rocks, peat or ice). Fossils include shells, hard skeletons, footprints, borings and casts.

fossil fuel Peat, coal, natural gas and petroleum - combustible material obtained from below ground, formed by the partial decomposition of plant and animal remains that were then subjected to varying degrees of pressure and heating. The energy stored in fossil fuels is derived originally from photosynthesis that occurred millions of years ago. Fossil fuel combustion releases gases, especially carbon dioxide, into the atmosphere. See *acid rain* and *carbon cycle*.

fovea A shallow pit in the *retina* of the vertebrate eye that contains a concentration of cones. In humans it is about 1 mm in diameter and is situated directly opposite the lens. It is the main area for acute and accurate vision in bright light.

frame quadrat An area in which organisms are sampled. The area is enclosed within a frame of any convenient shape, but usually square. See *quadrat*.

free energy A measure of the energy that becomes available to do useful work when released from a chemical reaction.

freezing The change of a liquid into a solid. See *water*.

fructose A simple hexose sugar occurring in green plants, fruit and honey. It is much sweeter than *sucrose* (formed when fructose combines with *glucose*).

fruit A plant structure consisting of the ripened ovary wall of a *flower*; it may or may not contain *seeds*. A fruit is either fleshy (succulent) or dry. Fleshy fruits are usually distributed by animals. They often have a brightly coloured outer layer (*epicarp*) that attracts animals, a thick and juicy middle layer (mesocarp), and an inner layer (*endocarp*) which may be fleshy or hard. Dry fruits are usually distributed by wind, water, or mechanical means. Those dry fruits which retain their seeds and do not break open are called *indehiscent*. Those which split open to release their seeds are called *dehiscent*.

functional gene See *cistron*.

Fungi A kingdom of *eukaryotes* that lack *chlorophyll* (and are therefore heterotrophic), but resemble plants in having rigid *cell walls* and being non-motile. Fungi may be microscopic (e.g. yeasts) or quite large (e.g. mushrooms and toadstools).

fungicide A chemical such as an organomercury compound or copper sulphate, that kills fungi. See also *pest*.

funicle See *ovule*.

gall bladder A sac-like structure, attached to the *bile* duct, which stores bile from the *liver* until this is needed in the *duodenum* to emulsify fats. Although present in most mammals, the gall bladder is absent in rats.

GALP See *glyceraldehyde 3-phosphate*.

gamete A cell that is involved in sexual reproduction. During *fertilization*, the haploid nuclei of two gametes fuse to form a diploid *zygote* capable of developing into a new individual. Most sexually reproducing organisms form two different types of gamete, one small and highly motile (the male gamete), the other larger and immobile (the female gamete), but some plants and fungi produce only one type of gamete (*isogametes*) all of which are similar in size and motility.

gametogenesis The formation of gametes. In mammals, *haploid* gametes are formed from *diploid* cells by two meiotic divisions (see *meiosis*), the egg cells in the ovary and the spermatozoa in the testis. In plants, gametogenesis usually occurs by *mitosis* in the *gametophyte* stage, although meiosis must occur at some stage of the life cycle, otherwise there would be a doubling of the *chromosome* number with each generation.

gametophyte The *haploid*, *gamete*-producing stage in the life cycle of plants. See also *alternation of generations*.

ganglion (*pl.* **ganglia**) A bundle of nerve cell bodies that in vertebrates is usually contained outside the *central nervous system*. Some masses of grey matter in the brain are also called ganglia. In invertebrates, ganglia form part of the central nervous system; for example, in earthworms, ganglia occur along the ventral nerve cord in each segment and in the head region.

gas exchange The process by which gases are exchanged between an organism and its external environment by *diffusion* across surfaces that are usually thin and have a large surface area to maximize the rate of exchange. It includes the uptake of oxygen and elimination of carbon dioxide in animal and plant *respiration* and the uptake of carbon dioxide and release of oxygen in plant *photosynthesis*.

gasohol Alcohols produced for fuel use. Ethanol is formed by the bacterial fermentation of organic matter, methanol is obtained from natural gas. Ethanol is produced commercially in fermenters, from maize in the US and sugar cane in Brazil, for use in road vehicles. See also *biogas*.

gastric juice The digestive juice secreted by the walls of the *stomach*. It includes hydrochloric acid, pepsin and rennin.

gastrin A *hormone* produced by the *stomach*, which stimulates the secretion of *gastric juices*. Gastrin secretion is stimulated by the presence of food in the stomach.

Gastropoda See *Mollusca*.

gene The fundamental unit of inheritance. A gene is the part of a *DNA* molecule involved in determining a particular characteristic (such as eye colour in humans) and which, in *eukaryotes*, occupies a specific position (locus) on a *chromosome*. A gene can exist in different forms (*alleles* or *allelomorphs*). Each allele consists of a unique sequence of nucleotides and may change the way in which a characteristic is expressed in the *phenotype* (for example, eye colour may be brown, green, or some other colour, depending on the particular alleles a person possesses). In diagrams used in genetics to display the results of particular crosses, the gene is often represented by a letter, with the upper case (capital) letter indicating the *dominant* allele and the lower case of the same letter indicating the *recessive* allele. In cases of *codominance*, the gene is allocated an upper case letter and each allele is allocated an appropriate superscript. For example, in humans the ABO blood group is determined by three alleles designated I^A, I^B, and I^O; I represents the gene locus and A, B and O represent alleles. The alleles I^A and I^B are codominant and the allele I^O is recessive to both I^A and I^B. This is a case of multiple alleles, since more than two alleles occur in the species, although only two can occur in any one diploid individual. People with blood group A have alleles $I^A I^A$ or $I^A I^O$; those with blood group B have alleles $I^B I^B$ or $I^B I^O$; those with blood group AB have alleles $I^A I^B$; and those with blood group O must have two recessive alleles $I^O I^O$. See also *genetic code*.

gene expression The manifestation in the *phenotype* of a particular *characteristic* specified by a *gene*. In introductory genetics it is often assumed that *dominant* alleles of a gene are always expressed fully, but this is not so. The expression in the phenotype of characteristics associated with one gene may be influenced by the presence or absence of another gene or genes (the genetic environment) and by the physical environment (temperature, pH, etc), and when a gene is expressed there are many possible degrees of intensity of expression.

gene mapping The construction of a diagram which shows the relative positions of *genes* on a particular *chromosome*. Gene mapping uses the cross-over value (COV). This is an estimate of the amount of *crossing-over* that occurs between two genes carried on the same chromosome. It is usually expressed as a percentage of the number of *recombinants* compared with the total number of offspring produced in a *dihybrid* cross involving linked genes:

% crossing over = (number of recombinants/total number of offspring) x 100.

Recombinants are offspring whose *phenotype* differs from that of either parent. It is assumed that the further apart two genes are on their chromosome, the more likely it is that the chromosome will break and rejoin at some point between them. By convention, a COV of 1% is taken to represent a distance of 1 unit on the chromosome.

gene mutation See *mutation*.

gene probe A specific fragment of single-stranded *DNA* (or *RNA*) which is usually labelled with radioactive phosphorus (^{32}P) to help in the detection of a gene with a complementary sequence of nucleotides to which the probe can bind by *base pairing*.

generation 1. A group of individuals of the same age or at the same stage of development in a population in which all members belong to the same species.
2. The time that elapses between the birth of parents and the birth of their offspring.
3. Production, especially of offspring.

generative nucleus One of two *haploid* male nuclei formed in the *pollen tube* of flowering plants. See also *double fertilization*.

generator potential The reversal of electrical charge (depolarization) that occurs when a receptor, *neurone* or muscle fibre is first stimulated. The stimulated part of the *cell membrane* becomes positively charged due to an influx of sodium ions. The generator potential is graded, the magnitude of the positive charge depending on the intensity of the stimulus; when this reaches a critical level (the threshold intensity) it triggers an *action potential*.

generic name See *binomial nomenclature*.

gene therapy The deliberate repair or replacement of damaged *genes* (e.g. those responsible for inherited diseases such as *cystic fibrosis*). So far, gene therapy has been used to repair somatic (body) cells rather than sex cells (gametes). It is generally agreed that it would be unwise to make inheritable modifications of sex cells until we have a better understanding of their effects.

genetic code The means by which the genetic information needed to make *proteins* is carried in *DNA*. According to the *one-gene-one-polypeptide theory*, a particular strand of DNA is thought to be responsible for the synthesis of a single *polypeptide* chain. The genetic code is based on the sequence of bases in the DNA strand. It is called a triplet code because each group of three bases (codon) codes for something very specific. As there are four different bases in DNA (i.e. it is a four-letter code), there are 64 possible codons. Most codons encode for specific *amino acids* in the polypeptide chain; there are 20 amino acids from which proteins are composed. The code is said to be degenerate, because it appears to contain more information than it needs and most of the amino acids are encoded by more than one codon. Some codons (called 'nonsense triplets') do not code for any amino acid, but signal the beginning or end of a particular polypeptide chain. Before protein synthesis takes place, the relevant piece of genetic code carried on DNA is transcribed (see *transcription*) into messenger RNA, which directs the translation of the code into the synthesis of polypeptides on *ribosomes* in the *cytoplasm*.

genetic engineering, genetic manipulation or **genetic modification** Modification of the genetic material of an organism for human benefit. Genetic engineering has recently become associated with the transfer of *DNA* from one organism to another; for example, a strand of human DNA responsible for insulin production can be transferred into bacterial cells (usually *Escherichia coli*) using *plasmids*; once incorporated into the bacterial *chromosome*, the human DNA can instruct the bacterium to produce human insulin.

genetic manipulation See *genetic engineering*.

genetic modification See *genetic engineering*.

genetics The scientific study of heredity and variation, and of how these are affected by the interaction between the environment and *genes*.

gene transfer The transfer of a strand of *DNA* carrying the *genetic code* for a particular *polypeptide* into an organism in which it does not naturally occur. Gene transfer forms the basis of most modern forms of genetic engineering.

genome The complete genetic constitution of an organism, a population of organisms, or a species. The term is sometimes used more specifically to refer to all the *genes* in a single set of *chromosomes* (i.e. in a haploid nucleus).

genotype The actual genetic constitution of an organism; usually, all the alleles which determine particular *characteristics*. Compare *phenotype*.

genus A group of organisms containing a number of closely related **species**. See *rank*.

geological time scale A time scale, based on the fossil record and other geological evidence, in which the history of the Earth is broken down into eons, eras, sub-eras, periods and epochs (for example, we are now living in the Holocene Epoch of the Pleistogene Period of the Quaternary Sub-era of the Cenozoic Era of the Phanerozoic Eon). See appendix 2.

geotaxis See *taxis*.

geotropism See *tropism*.

germination The growth of a seed, spore, or plant structure (e.g. *pollen tube*) usually following a period of dormancy. In seeds, germination is the onset of growth of the *embryo* and requires water, oxygen, and a suitable temperature. Some seeds require other specific conditions before they will germinate (e.g. exposure to light, freezing temperatures, fire or drought). In all types of germination, the seed rapidly takes up water which is used for the activation of *enzymes* and *hydrolysis* of insoluble food stores into soluble substances. These substances are readily transported to different parts of the germinating seed and are used as an energy source or as raw material for growth. Seed germination is completed when the young plant is able to manufacture its own food by *photosynthesis*.

gestation period See *pregnancy*.

gibberellins See *plant growth substance*.

gills 1. Filamentous outgrowths of the body wall or the wall of the *pharynx*, which act as the respiratory organ of aquatic animals. Fish gills provide a large surface area, are very thin, and are well supplied with blood, so *gas exchange* is efficient. **2.** Plate-like, spore-producing structures that radiate from the underside of the umbrella-shaped cap of agaric fungi (e.g. mushrooms and toadstools).

global warming See *greenhouse effect*.

globular protein A *protein* that has at least one highly folded polypeptide chain. Globular proteins have a roughly spherical shape and are relatively easy to combine with water to form *colloids*. *Enzymes*, *antibodies*, and some *hormones* are globular proteins.

glomerular filtrate See *glomerulus*.

glomerulus A knot of *capillaries* in the Bowman's capsule (a cup-shaped structure formed at the top of a *nephron*). Blood, entering the glomerulus under high pressure, is filtered through the basement membrane and enters the nephron as the glomerular filtrate, consisting of water, its dissolved solutes (e.g. glucose and urea), and other small molecules from the blood.

glucagon A *hormone*, secreted by the alpha cells of the *islets of Langerhans* (see *pancreas*), that promotes the breakdown of *glycogen* to *glucose*; its actions oppose those of *insulin*.

glucose A hexose (six-carbon) monosaccharide sugar which has the chemical formula $C_6H_{12}O_6$. In nature, glucose usually forms one of two types of ring structure, called alpha-glucose and beta-glucose. Alpha-glucose molecules combine to form *starch* in plants and *glycogen* in animals; beta-glucose molecules combine to form *cellulose*. Glucose plays a vital role in the *metabolism* of organisms: it is a major energy source, being a *substrate* for both *aerobic* and *anaerobic respiration*; and it is a major product of photosynthesis. See diagram overleaf.

glyceraldehyde 3-phosphate (GALP) or **phosphoglyceraldehyde (PGAL)** A three-carbon (triose) sugar phosphate which plays an intermediate role in both *respiration* and photosynthesis. In respiration, two molecules of GALP are formed during *glycolysis* by the breakdown of one molecule of fructose 1,6 diphosphate. GALP is then broken down in several steps to *pyruvate*. In photosynthesis, GALP is formed after the reduction of glycerate-3-phosphate by NADPH. Some of the GALP is used to regenerate *ribulose biphosphate* and some to synthesize *glucose*.

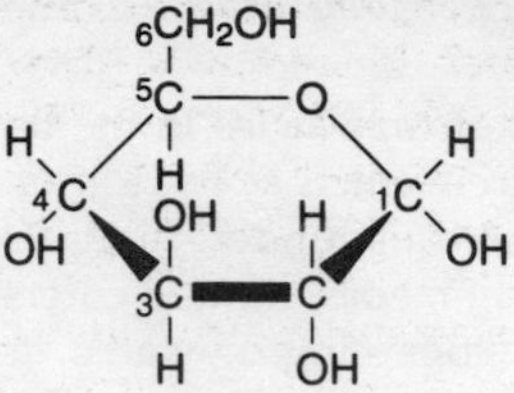

α-glucose

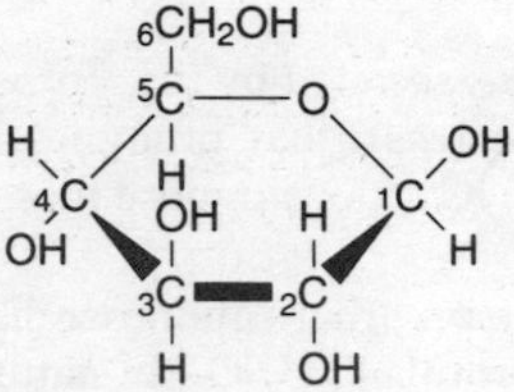

β-glucose

glucose

glycerate 3-phosphate (GP) or **phosphoglyceric acid (PGA)**
A three-carbon molecule produced from *glyceraldehyde
3-phosphate* (GALP) during *glycolysis*. It also plays an
important role in the *Calvin cycle* of photosynthesis, being the
first stable product formed after fixation, when carbon dioxide
combines with *ribulose biphosphate*.

glycerol An alcohol derived from a three-carbon sugar. It is a
component of most naturally occurring *fats* and *oils* and is
released in the gut when these are digested.

glycocalyx See *fluid-mosaic model*.

glycogen or **animal starch** A highly branched polysaccharide
composed of alpha-glucose molecules (see *glucose*). It is the

main *carbohydrate* storage product of animals. Chemically, it is very similar to *starch*. In mammals, glycogen occurs mainly as tiny granules in muscle and liver cells. It is easily hydrolysed (see *hydrolysis*) into glucose and acts as a readily available source of energy for metabolically active cells.

glycolysis The first stage of *respiration*, which takes place in the *cytoplasm* with or without oxygen. One molecule of *glucose* is broken down to two molecules of *pyruvic acid*. During the process, *NAD* is reduced, generating energy used to make four molecules of *ATP*. Two ATP molecules are required as activation energy for glycolysis to take place, so the process results in a net production of two molecules of ATP from each molecule of glucose.

glycoprotein Any protein that contains a sugar as part of the molecule. Glycoproteins form part of the glycocalyx of cell surface membranes (see *fluid-mosaic model*).

glycosidic bond The oxygen-containing bond formed between two sugar monomers following a *condensation* reaction. For example, glycosidic bonds link together the *glucose* monomers in *starch*, *cellulose*, *glycogen*, and *maltose*.

goblet cell A wine-glass-shaped gland cell that secretes mucus. Goblet cells are abundant in the *epithelium* lining the small *intestine* and the *trachea*.

goitre See *thyroid gland*.

Golgi apparatus Stacks of flattened, membrane-bounded sacs (*cisternae*) present in cells of *eukaryotes*; each stack is called a *Golgi body*. The Golgi apparatus processes, stores, packages and transports substances. It directs secretory lipids and proteins to their correct destination in the cell. It is also involved in *lysosome* formation.

Golgi body See *Golgi apparatus*.

gonad A reproductive organ that produces the *gametes*. The ovaries and testes are the gonads in mammals, producing egg cells and spermatozoa respectively, and also secreting hormones.

gonorrhoea A disease in humans, caused by the bacterium *Neisseria gonorrhoea* that affects the mucus lining of the male and female reproductive organs, usually producing painful discharges in the penis and vagina. Treatment in the early stages (with e.g. penicillin) is usually effective. If untreated, the disease can lead to dangerous secondary complications, including disorders of the heart and eyes. Gonorrhoea is transmitted during sexual intercourse and is highly contagious.

GP See *glycerate 3-phosphate*.

Graafian follicle A large, fluid-filled ball of cells in the *ovary* that contains an *oocyte* attached to its wall. When the oocyte is released during *ovulation*, the Graffian follicle develops into a temporary *endocrine gland*, called the *corpus luteum*.

Gram staining A procedure (first devised by Christian Gram) for differentiating bacteria that may have a similar shape. It consists of staining a heat-fixed smear of bacteria with crystal violet for 30 seconds, rinsing the stain off with Gram's iodine solution, and then treating with ethanol. Bacteria decolorized by the ethanol are called Gram-negative, and those which retain the blue—purple dye are called Gram-positive. (Differences in stain retention are related to differences in cell wall structure.) Another stain is then applied which is taken up by Gram-negative bacteria but not Gram-positive ones.

grana See *chloroplast* and *thylakoid*.

granulocyte A *white blood cell* that contains granules in its *cytoplasm*.

graph A diagram, usually showing the relationship between two variables plotted on axes at right angles to each other. Various types of graph are suitable for the presentation of different data (see appendix 1).

grazing Feeding on vegetation, usually grass, that is close to the ground. Grazing by sheep or other herbivorous animals is often used in grassland management, to prevent grassland from developing into scrub or woodland by natural *succession*. Grazing tends to suppress *dicots* (trees and shrubs), in which the growth areas are at the tips of the shoots, but not grasses, in which the growth areas are at the base of the shoots.

greenhouse effect The trapping of heat in the Earth's atmosphere by certain gases (especially water vapour, carbon dioxide, methane and chlorofluorocarbons), which act in a manner similar to glass in a greenhouse. They are transparent to shortwave radiation, but absorb and reradiate some of the longwave radiation (heat) from the Earth's surface. The greenhouse effect is not new. Without it the present average temperature at the

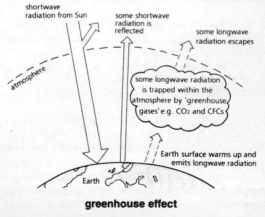

shortwave radiation from Sun

some shortwave radiation is reflected

some longwave radiation escapes

atmosphere

some longwave radiation is trapped within the atmosphere by 'greenhouse' gases' e.g. CO_2 and CFCs

Earth surface warms up and emits longwave radiation

Earth

greenhouse effect

Earth's surface would be −23 °C (rather than the actual value of +15 °C), which might be too low for life as we know it to have evolved. There is considerable anxiety about the recent increase in carbon dioxide and other so-called 'greenhouse' gases, however, because they may cause an undesirable increase in the average surface temperature (*global warming*).

grey matter Areas in the *central nervous system* where there is a high density of of cell bodies. Grey matter occurs in the central area of the spinal cord and on the surface of the cerebral cortex, where much of the processing of information takes place.

gristle See *cartilage*.

gross productivity See *productivity*.

ground tissue Plant tissue other than the *epidermis*, reproductive tissues, and *vascular tissues*. It is formed from the apical *meristem* and consists mainly of *parenchyma*, but may also include *collenchyma* and *sclerenchyma*.

growth A relatively permanent increase in the size of an organism. It may be measured by increases in linear dimensions, but is better measured in terms of dry weight, because temporary changes due to intake of water are not regarded as growth. Growth of a multicellular organism usually occurs in three phases: cell division; cell assimilation; and cell expansion.

growth curve A *graph* showing changes in the size of a population, organism, or part of an organism. A growth parameter is plotted using the y-axis and time is plotted using the x-axis. There are many types of growth curve, but it is commonly sigmoid (S-shaped) and has three phases: an initial phase of slow growth (lag phase); a central rapid phase showing exponential growth (log phase); and a final period of slow growth.

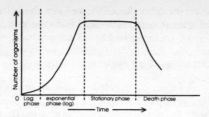

growth curve Typical growth curve of a bacterial population.

guanine See *purine*.

guard cell See *stoma*.

gustation See *taste*.

gut The *alimentary canal* or part of it.

guttation See *root pressure*.

Gymnospermae See *Coniferophyta*.

gynaecium See *flower*.

habitat The particular locality in which an organism lives. Some organisms occupy fairly large areas and their habitat description need not be very precise (e.g. deciduous oak woodland or upper part of rocky shore), but small organisms often occupy small areas (*microhabitats*) with very specific conditions. The *microclimate* (the precise atmospheric conditions, including air temperature, intensity of sunlight, and relative humidity) of two adjacent microhabitats (e.g. the upper and lower sides of a boulder) may be very different.

habitat niche See *niche*.

137

habituation A form of learned behaviour in which the response to a stimulus is reduced if the stimulus is repeated constantly.

haemocoel See *circulatory system*.

haemoglobin A large, conjugated, *globular protein* molecule (see *conjugated protein*) consisting of four polypeptide chains each attached to an iron-containing *prosthetic group*. Haemoglobin is found in the blood of all vertebrates (animals with backbones) and many invertebrates (animals without backbones). Each haemoglobin molecule can pick up, transport, and unload up to four oxygen molecules. It also occurs in the root nodules of leguminous plants (i.e. those of the pea family), but only if they contain *Rhizobium* (a nitrogen-fixing bacterium).

haemolysis The breakdown of red blood cells. If, for example, red blood cells are immersed in distilled water, the water enters by *osmosis* and inflates the cells until they burst. Haemolysis may also result from disease, poisons, or incompatible blood transfusions.

haemophilia A genetic disorder in humans, caused by a defective allele which results in the lack of a clotting factor (either Factor VIII or Factor IX). The disorder is characterized by excessive bleeding and an inability of the blood to clot normally. Haemophilia exhibits sex-linkage because the allele is **recessive** and carried on the X-chromosome; it therefore occurs mainly in males.

hair 1. An epidermal outgrowth of mammalian skin made from dead cells containing the protein *keratin*. Hair has many functions, including thermal insulation, protection against harmful radiation from the Sun, and camouflage. Nerve endings close to hairs give them a sensory function.
2. See *root hair*.

halophyte A terrestrial plant that can thrive in soils with a high concentration of salts (e.g. on salt marshes or at the top of a beach only occasionally covered by sea water). Most halophytes actively transport salts into their root cells to create a *water potential* lower than that in the surrounding soil; water then passes into the roots by *osmosis*. Some halophytes, e.g. cord grass (*Spartina*) eliminate excess salts from special glands at the margins of their leaves. Compare *hydrophyte, mesophyte* and *xerophyte*.

hammer See *malleus*.

haploidy The state of having body cells each of which contains one set of *chromosomes* (i.e. the haploid state). In some organisms (e.g. *bryophytes* and some *fungi*) the most persistent and conspicuous stage of the life cycle is haploid. If a *mutation* produces a **recessive** allele, this will be immediately expressed and exposed to *natural selection*. Compare *diploidy*.

Hardy—Weinberg equation An equation used to determine the frequency of *genotypes* within a population. If a gene has two alleles, A and a with a frequency of p and q respectively, the frequency of genotypes will have the proportions: $p^2 + 2pq + q^2 = 1.0$, where p^2 represents the homozygous *dominants*; $2pq$ the *heterozygotes*, and q^2 the homozygous *recessives*.) Since $p + q = 1$, it is simple to determine all the genotype frequencies from the proportions of the recessive phenotypes. G.H. Hardy and W. Weinberg independently demonstrated in about 1908 that the allele frequencies remain relatively constant provided that the population is large; immigration = emigration; deaths = births; there is no *natural selection* or *mutation*; and mating is random.

health The absence of disease or injury. In humans, health may also be defined positively, as the ability to mobilize all physical, mental, and spiritual resources for the benefit of the individual and the society to which she or he belongs.

heart A muscular part of the *circulatory system* in animals, that contains valves to prevent back-flow. The mammalian heart consists of four chambers: right *atrium*; right *ventricle*; left atrium; and left ventricle. The right and left sides are separated by a *septum* (dividing wall) and function as two separate pumps. The right side pumps deoxygenated blood to the lungs; the left pumps oxygenated blood to the rest of the body. Deoxygenated blood returns to the heart through the *vena cava* and into the right atrium; the superior vena cava carries blood from the head, neck and upper limbs, the inferior vena cava carries blood from the rest of the body. The atria pump the blood simultaneously to the right and left ventricles, which then pump the blood to the lungs and rest of the body respectively. The wall of the left ventricle is the more muscular because it must pump blood furthest. The heart has its own inherent rhythm (about 72 beats per minute), maintained by a

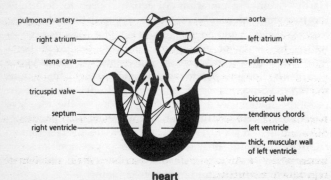

heart

pacemaker region and modified by *hormones* and nervous stimulation. The unidirectional blood flow is maintained by valves. A *tricuspid valve* on the right side and a *bicuspid valve* on the left side prevent back-flow from the ventricles to the atria, and *semilunar valves* in the pulmonary artery and dorsal *aorta* prevent back-flow into the heart from these vessels. See also *cardiac cycle*.

heart beat See *cardiac cycle*.

heart muscle See *cardiac muscle*.

hepatic Pertaining to the *liver* or its functions.

hepatic portal vein The *blood vessel*, in mammals, that carries deoxygenated blood from the small *intestine* to the *liver*, where substances absorbed from the *gut* are processed.

herb See *herbaceous plant*.

herbaceous plant or **herb** A flowering plant that has little permanent woody tissue in its stem or roots. All the aerial parts of herbs die back at the end of the growing season.

herbivore Any animal that feeds on plant material. Herbivores must break open the *cellulose* cell walls of plants to obtain nourishment. Herbivorous mammals (e.g. sheep and cattle) have *molar* teeth adapted to grind plant material into very fine pieces. Symbiotic bacteria, which digest the cellulose, inhabit the *alimentary canal*.

heredity The transfer of genetic information from parents to offspring.

hermaphrodite An organism that has both male and female reproductive structures.

heroin A white, crystalline substance, derived from morphine, which acts as a narcotic, inducing sleep. It has no medical uses and is highly addictive.

heterochromatin See *chromatin*.

heterodont dentition See *tooth*.

heterogametic See *sex determination*.

heterosomal inheritance The inheritance, in a diploid organism (e.g. a human), of *characteristics* determined by *genes* which are carried on heterosomes (i.e. sex chromosomes which are not identical in appearance). Compare *autosomal inheritance*.

heterosome See *heterosomal inheritance*.

heterospory A reproductive strategy in which plants produce two different types of *spore* — *megaspores* and *microspores* — which give rise to distinct female and male *gametophytes* respectively. All conifers and flowering plants exhibit heterospory. The development of heterospory is thought to have been an important step in the evolution of seed-bearing plants. Compare *homospory*.

heterotroph See *heterotrophic nutrition*.

heterotrophic nutrition Nutrition in which an organism (a *heterotroph*) obtains complex organic compounds from other organisms. All animals, all fungi, and most bacteria are heterotrophs, including those bacteria (called *photoheterotrophs*) which use sunlight to synthesize organic molecules from organic raw materials obtained from other organisms. Heterotrophic nutrition includes *holozoic* nutrition, parasitic nutrition (see *parasite*), symbiotic nutrition (see *symbiosis*), and *saprobiontic nutrition*.

heterozygote A diploid cell or organism (see *diploidy*) in which two different alleles of a particular *gene* occur at the same locus on *homologous chromosomes*.

Hexapoda See *Insecta*.

hibernation The passing of the winter by an animal in a torpid state, characterized by a reduced metabolic rate and significant lowering of body temperature. It is a strategy for coping with long periods of cold weather when food may be scarce.

hip girdle See *pelvic girdle*.

Hirudinea See *Annelida*.

histogram See appendix 1.

histology The microscopical study of tissues.

histone A simple *protein* that is closely associated with *DNA* in the *chromosomes* of *eukaryotes*. Portions of the DNA helix appear to wrap themselves round groups of eight histone molecules, forming structures called nucleosomes. Histones are thought to be involved in the condensation and coiling of chromosomes during *mitosis* and *meiosis*. They may also regulate the activity of DNA in some way.

HIV See *AIDS*.

holophytic nutrition *Autotrophic nutrition* carried out by *photoautotrophs*, especially green plants.

holozoic (of heterotrophs) Feeding on solid organic matter derived from other organisms (see *heterotrophic nutrition*).

homeostasis or **homoeostasis** The regulation and maintenance of relatively constant conditions within a system. It is a basic feature of all living things, occurring at every level of organization, from the cell to the *ecosystem* and, according to the 'Gaia hypothesis' of James Lovelock, even to the *biosphere*. Physiological homeostasis is the maintenance of constant conditions in the *internal environment* of an organism. In mammals, this involves negative *feedback* mechanisms controlling water volume, *blood glucose* content, and temperature.

homeotherm or **homoiotherm** An organism that maintains a relatively constant body temperature despite fluctuations in environmental temperature. It may achieve this by internal, metabolic processes (see *endotherm*) or behavioural mechanisms (see *ectotherm*). Compare *poikilotherm*.

homodont dentition See *tooth*.

homoeostasis See *homeostasis*.

homogametic See *sex determination*.

homogenate See *subcellular fractionation*.

homogenizer A device that breaks open cells and releases their *organelles*. Homogenizers vary in design, but typically work much like a food mixer: as they rotate, very sharp blades, set at right angles to each other, produce powerful shearing forces that cut open the cells.

homoiotherm See *homeotherm*.

homologous chromosomes Pairs of *chromosomes* in a diploid organism (see *diploidy*), which usually appear similar and carry the same *genes*, although the alleles on each chromosome may differ. Homologous chromosomes pair during *meiosis*.

homospory A reproductive strategy in which plants produce only one type of *spore* that gives rise to a *hermaphrodite* gametophyte generation. Homospory occurs in all *bryophytes* and some *ferns*. Compare *heterospory*.

homozygosity See *inbreeding*.

homozygote A diploid cell or organism (see *diploidy*) in which two identical alleles of a particular *gene* occur at the same locus on *homologous chromosomes*.

hormone A chemical produced in one part of the body and transported to another part where it produces specific effects. Animal hormones are produced by *endocrine glands* and secreted, usually in very small amounts, directly into the blood stream, which carries them to their target organs. See also *plant growth substance*.

horticulture See *agriculture*.

host An organism in or on which another organism lives for the whole or part of its life and from which it gains some benefit (e.g. nourishment, shelter, or transport).

human being The common name for the mammalian species *Homo sapiens*. Humans are primates, closely related to the chimpanzees, but are the only existing species in the *genus Homo*.

human growth hormone A *hormone* secreted by the anterior *pituitary* gland under the influence of the hypothalamus (see *brain*). It promotes cell growth, protein synthesis and, in children, elongation of the long bones. In children, deficiencies can lead to dwarfism and excesses to gigantism. Human growth hormone manufactured by *genetic engineering* has been used to treat children with hormone deficiency and, more contro-versially, to increase the height of otherwise normal children.

human immunodeficiency virus See *AIDS*.

humerus The bone in the upper arm or upper forelimb of land vertebrates.

humus A complex, gelatinous, organic material that gives soil its dark brown colour. Humus is formed from the decomposi-tion of organisms but has lost all visible trace of the living material from which it is derived. Its presence confers a number of advantages on soils: it acts as a *colloid*, retaining moisture and binding mineral fragments together; it improves soil texture, making it more friable and easier to work; it absorbs solar radiation, enabling soil to warm quickly, thus promoting seed germination and plant growth; and the nutrients it con-tains support many soil organisms, including earthworms and bacteria, which play an important role in *nutrient recycling*.

hybrid Any offspring of a cross between two genetically dissimilar individuals. See also *heterozygote*.

hybridoma See *monoclonal antibody*.

hydathode See *root pressure*.

hydrocarbon chain Part of an organic molecule that consists of a chain of hydrogen and carbon atoms.

hydrogen acceptor or **hydrogen carrier** A molecule that readily accepts a hydrogen atom. In the light stage of photo-synthesis, for example, *NADP* acts as a hydrogen acceptor, combining with the hydrogen made available by the *photo-lysis* of water. In *aerobic respiration*, *NAD* acts as the first of a series of hydrogen acceptors (a hydrogen chain) in the respiratory chain. Hydrogen acceptors readily gain hydrogen (reduction) and lose hydrogen (oxidation). These redox reactions are important for generating energy used for *ATP* synthesis during *respiration* and photosynthesis.

hydrogen bonding The weak electrostatic bonding that occurs between a slightly positive hydrogen atom and another atom with a slight negative charge. Hydrogen bonds form between hydrogen and oxygen atoms in water molecules, and between —OH or —NH groups and —C=O groups in other molecules. Hydrogen bonds are important for stabilizing the structure of some molecules, including *cellulose*, *DNA* and some *enzymes*.

hydrogen carbonate indicator A solution that indicates the occurrence of *respiration* and photosynthesis. At a neutral *pH*, the indicator is red. It changes to a yellowish colour when it becomes acidic in the presence of carbon dioxide produced during respiration, and to a deep purple colour when it becomes more alkaline as a result of carbon dioxide being extracted and oxygen eliminated during photosynthesis.

hydrogen carrier See *hydrogen acceptor*.

hydrogen chain A series of *hydrogen acceptors* that form the first part of the respiratory chain in *aerobic respiration*.

hydrogen ion concentration See *pH*.

hydrolase or **hydrolytic enzyme** An *enzyme* that catalyses *hydrolysis* reactions. Digestive enzymes are hydrolases.

hydrological cycle See *water cycle*.

hydrolysis A chemical reaction that results in the breakdown of a large molecule by the addition of water. Compare *condensation*.

hydrolytic enzyme See *hydrolase*.

hydrophyte A flowering plant that is adapted to live in freshwater or very wet conditions. Hydrophytes include the Canadian pondweed (*Elodea canadensis*) and water-lilies (*Nymphaea*). The leaves of hydrophytes are usually submerged or floating and possess *stomata* only on aerial surfaces. Roots, vascular tissue and mechanical-support tissue are usually reduced or absent; leaves and stems are often filled with air to aid flotation; and submerged parts lack a *cuticle*. Compare *halophyte, mesophyte* and *xerophyte*.

hydrostatic skeleton A watery fluid that forms a rigid medium (because of the relative incompressibility of water) against which muscles can contract. Hydrostatic skeletons are found in soft-bodied organisms (e.g. earthworm) or soft structures (e.g. penis). The incompressible fluid in the large *vacuoles* of plant cells also function as a hydrostatic skeleton. The mechanical strength of fully turgid cells is sufficient to support small plants such as mosses. See also *coelom*.

hydrotropism See *tropism*.

hymen A mucus membrane stretched across the *vagina* in some female mammals. In humans, the hymen usually contains a small opening and is broken open during the first act of sexual intercourse.

hyperglycaemia The presence of an abnormally high concentration of *blood glucose*. It is a symptom of *diabetes mellitus*, but may also occur temporarily after a meal rich in sugars.

hypermetropia See *long sightedness*.

hypertension See *blood pressure*.

hyperthyroidism See *thyroid gland*.

hypertonic (of a solution) Having, usually, a lower *water potential* than another fluid, in e.g. an animal cell, thus causing a cell immersed in it to shrink as water is lost from the cell by *osmosis*.

hypha (*pl.* **hyphae**) A branched, thread-like, tubular structure which forms the basic unit of the vegetative growth stage of *fungi* such as mushrooms and toadstools. The mass of hyphae which make up the bulk of a fungus is called a mycelium. Each hypha is composed of a *cell wall* usually containing *chitin* and a layer of *cytoplasm* surrounding a *vacuole*. The tubes may be continuous and multinucleated, partially divided, or completely subdivided along their length into compartments each with one or two nuclei. *Enzymes* are secreted from the tips of the hyphae to digest the substrate in which they are growing.

hypothyroidism See *thyroid gland*.

hypoglycaemia Low *blood glucose* concentration; it may be due to lack of food, over-exertion, or overproduction of *insulin*.

hypothalamus See *brain*.

hypothesis An explanation of scientific observations or phenomena. An hypothesis should be stated in such a way that it can be tested by an experiment which either contradicts (refutes) or supports it.

hypotonic (of a solution) Having (usually) a higher *water potential* than another fluid, in e.g. an animal cell, causing water to enter the cell by *osmosis*, resulting in an increase in the volume of the cell. Certain solutions with the same water potential as the cell can also be hypotonic. For example, if *red blood cells* are immersed in a solution of *urea* they will burst because water molecules follow urea molecules that can diffuse into the cell.

IAA See *plant growth substance*.

I band See *sarcomere*.

ice formation See *water*.

ileum The final part of the small *intestine* in mammals, lying between the *duodenum* and *colon*. It is the region in which digestion of fats, carbohydrates and proteins is completed. The epithelial lining of the ileum is similar to that of the duodenum; it has *crypts of Lieberkuhn*, *goblet cells* and *enzyme*-secreting cells, is highly folded, and has *villi* which provide a large surface area for the absorption of digested products. Unlike the duodenum, however, the ileum has no Brunner's glands.

imago The adult, sexually mature stage in the life cycle of a *pterygote* insect (i.e. a winged insect or an insect which has evolved from a winged insect).

immobilization The restriction of the movement of cells or enzymes, in *biotechnology*, by attaching them to the surface or in the spaces of plastic foam, beads or sheets. As with suspensions, the cells and enzymes are in close contact with the circulating medium, but immobilization allows them to be manipulated easily. For example, enzymes can be separated from their product and re-used.

immunization A method of rendering an animal more resistant to harmful microorganisms and their products by artificial means. Active immunization involves the introduction of *antigens* in the form of a vaccine containing dead or inactivated *bacteria* or *viruses*; the antigens stimulate the immune system to produce specific *antibodies* which protect the animal against future exposure to the antigen. Passive immunization involves the administration of a serum containing preformed antibodies.

immunoglobulin One of a group of highly variable proteins made by blood *lymphocytes*, which has *antibody* activity.

implantation The attachment of a fertilized egg cell to the *uterus* in mammals.

imprinting A form of learning in which a young animal learns to direct some of its social responses to a particular object, usually its parent. It was originally proposed by Konrad Lorenz to explain the behaviour of certain young birds (e.g. geese), that follow a moving object to which they are exposed during a brief, critical period in their lives. The behaviour is irreversible and influences future patterns of social behaviour (e.g. sexual behaviour).

inbreeding Mating between closely related individuals, or *self-fertilization*. Inbreeding tends to reduce genetic variability and increase homozygosity (the presence in a diploid organism of pairs of identical alleles for one or more *genes*). Consequently, inbreeding tends to increase the incidence of recessive *phenotypes* which may lead to a lowering of viability. Compare *outbreeding*.

incipient plasmolysis See *plasmolysis*.

incisor One of the chisel-shaped teeth in the front of the jaw in mammals, used for cutting, gnawing, and grooming. In some herbivores (e.g. rabbits) the incisors continue to grow throughout life.

incus or **anvil** A small, anvil-shaped bone that forms one of the three *ear ossicles* which transmit sound vibrations through the middle *ear* to the inner ear.

indehiscent See *fruit*.

independent assortment See *law of independent assortment*.

indicator organism An organism that survives only within a restricted range of a particular environmental factor or set of factors, so information about the levels of that factor can be derived from its presence and abundance. Examination of lichens within an area can provide information about the level of atmospheric pollutants (e.g. sulphur dioxide and phosphate), for example, because lichen species vary in their ability to tolerate these and, being long-lived, their growth and survival is influenced over a long period and may provide a better assessment of pollution than accurate measurements of concentrations taken on a particular day.

indoleacetic acid See *plant growth substance*.

induced-fit model See *lock-and-key theory*.

infection The invasion of the body of any living organism by a harmful microorganism (pathogen) which can reproduce within the host and produce harmful symptoms.

ingestion The taking of complex organic food into the body of an organism. In multicellular animals, ingested food enters the *alimentary canal*, where it is digested and absorbed.

inhalation or **inspiration** The act of drawing air or water into the body of an animal so *gas exchange* can take place in the respiratory organ (e.g. trachea of insects, gills of fish, lungs of mammals).

inheritance The acquisition of *characteristics* by the transfer of *genes* from parent to offspring.

inheritance of acquired characteristics See *Lamarckism*.

inhibition A process that reduces the effectiveness of an *enzyme*. An inhibitor reduces, or even stops, the ability of an enzyme to catalyse a reaction. Competitive inhibition involves an inhibitor that closely resembles the shape of the *substrate*; the inhibitor competes with the substrate for the *active site*. The amount of inhibition depends on the affinity between the enzyme and substrate and between the enzyme and inhibitor, and on the relative concentrations of the inhibitor, substrate and enzyme. Competitive inhibition is reversible. Non-competitive inhibition occurs when the inhibitor does not resemble the shape of the substrate. Some heavy metals are non-competitive inhibitors that attach to the active site, altering its shape and making it incapable of combining with a substrate. *Allosteric inhibitors* occur on special enzymes (allosteric enzymes) that have two specific attachment sites: an active site for a substrate and a separate site for an inhibitor.

innate (of **behaviour**) Not learned (i.e. it is inherited genetically).

inner ear See *ear*.

inoculation The introduction of biological material (the *inoculum*), consisting of microorganisms or cells from a multicellular organism, into a new medium (e.g. another organism, an artificial nutrient substrate, or soil).

inorganic Relating to the chemistry of all elements except carbon, and their compounds.

Insecta, Hexapoda or **insects** A class of mainly terrestrial animals (phylum *Arthropoda*) in which adults are usually winged (although the wings may have been lost in the course of evolution) and have a well-developed head region with compound eyes. The body is divided into a head, thorax and abdomen. Three pairs of jointed legs occur on the thorax.

insecticide A chemical (e.g. *DDT*) that kills insects. See also **pest**.

insects See *Insecta*.

insertion See *mutation*.

insight learning A form of learning in which an animal solves a problem in a situation it has not experienced before by using previously learned behaviour to produce a new response, rather than by trial and error. This appears to involve reasoning.

inspiration See *inhalation*.

instinct Behaviour that is believed to be completely inherited (i.e. under genetic control) and which is not modified by learning. The term is little used by ethologists (scientists who study and describe animal behaviour), because it is now known that most behaviour consists of a combination of learned and innate (inherited) components. For example, bird song was once thought to be entirely instinctive, but it is now known that, in some birds, the chick modifies the song even before it hatches from the egg.

insulin A polypeptide *hormone* secreted by beta cells in the islets of Langerhans of the *pancreas*. It promotes the conversion of *glucose* to *glycogen* in the *liver* and the uptake of glucose by cells in other parts of the body. Insulin deficiency causes high *blood glucose* levels and is one cause of *diabetes mellitus*. Insulin is used to treat diabetes; it was the first human substance to be produced commercially by *genetic engineering*.

integrated control See *pest*.

integrated pest management See *pest*.

integument The surface covering of an animal; examples are skin and cuticle.

intelligence The ability to adapt to new situations in a purposeful manner and to understand and grasp abstract concepts. There is no universally agreed measure of intelligence and intelligence tests in humans have been notoriously unreliable.

intensive farming The cultivation of crops and rearing of farm animals using highly technological methods (e.g. the heavy use of pesticides and *fertilizers* and extensive use of machinery) to obtain the maximum production from an area of land. Factory farming, where animals are kept indoors under tightly controlled environmental conditions and fed artificial foods, is an extreme form of intensive farming.

intercellular matrix See *connective tissue*.

intercostals In mammals, two sets of *antagonistic muscles* between the ribs that help to ventilate the lungs. Contraction of the external intercostals moves the ribs upwards and outwards, increasing the volume of the lungs, reducing the pressure, and sucking air inwards. Contraction of the internal intercostals has the opposite action.

internal environment The environment immediately surrounding the cells of a multicellular organism; the *tissue fluid*. See also *homeostasis*.

interneurone See *reflex action*.

internode The part of the plant stem between the nodes (the parts of the stem from which leaves grow).

interoceptor See *receptor cell*.

interphase A growth stage in the *cell cycle* during which the *chromosomes* are in a more or less dispersed state (see *chromatin*). Three subdivisions of interphase have been identified: G1, during which the cell is rapidly growing and metabolically active, and the *centrioles* replicate; S, during which the cell continues to grow and *DNA* is replicated in preparation for division of the nucleus; and G2, during which the cell continues to grow, and the proteins of the *spindle apparatus* are organized. During interphase, *organelles* are also replicated and *ATP* stored for use in nuclear division.

interspecific competition See *competition*.

intervertebral disc See *vertebra*.

intestine The region of the *alimentary canal* between the *stomach* and the *anus* or *cloaca*. In vertebrates, it is usually divided into a small intestine, concerned with the digestion of food and absorption of digested products, and a large intestine, concerned mainly with reabsorption of water and formation of *faeces*.

intine See *pollen grain*.

intracellular digestion See *digestion*.

intraspecific competition See *competition*.

inversion See *mutation*.

invertase See *sucrase*.

in vitro (*Latin* — in glass) (of biological processes) Occurring under experimental conditions outside a cell or organism, usually in a test tube or other glassware.

in vivo (*Latin* — in life) (of biological processes) Occurring under natural environmental conditions inside a cell or inside a living organism.

involuntary muscle See *smooth muscle*.

iodine An element required in trace quantities for healthy growth and development. Iodine is a constituent of thyroxine, the *hormone* secreted by the thyroid gland that controls the rate of *metabolism*. In mammals, iodine deficiency results in an underactive thyroid gland; the gland becomes swollen (a condition known as goitre), metabolism slows down and the animal becomes sluggish.

iodine test See *starch test*.

ion A charged particle produced when an atom either loses an **electron** to become positively charged (cation), or gains an electron to become negatively charged (anion).

IPM See *pest*.

iris See *eye*.

iron An element that is essential for both plants and animals. It is a component of some *electron carriers*, enabling them to carry out redox reactions in photosynthesis and *respiration*; it is an activator (see *cofactor*) of *catalase*; it is required for *chlorophyll* synthesis, and it is an important part of *haemoglobin*. Iron deficiency causes anaemia in mammals. The best source of iron for humans is red meat, especially liver, in which the iron is within haemoglobin (haem iron).

irrigation 1. The supplying of water to agricultural land by artificial methods (e.g. by spraying or diverting natural water courses into channels).
2. A technique used in light microscopy for staining and washing temporary sections of tissues on slides. The section is placed on a slide in a drop of water (or stain) and covered with a coverslip. Water (or stain) is dropped onto the edge of one side of the coverslip and filter paper placed against the opposite edge. Water (or stain) is drawn across the section by *capillarity* and is soaked up by the filter paper.

irritability or **sensitivity** The ability of an organism to respond to a stimulus in a way that tends to improve its chances of survival. Irritability is one of the fundamental features of living things.

islets of Langerhans See *pancreas*.

isoelectric point See *zwitterion*.

isogamete See *gamete*.

isolating mechanisms Factors that prevent breeding between two populations and lead eventually to speciation. Isolating mechanisms include physical barriers (e.g. mountains, rivers and seas), resulting in geographical isolation; differences in the ways organisms relate to their environment, resulting in ecological isolation; differences in breeding behaviour or the timing of breeding cycles, resulting in reproductive isolation; anatomical differences that prevent mating; or genetic incompatability between two populations that can mate but either produce no offspring or produce sterile offspring, resulting in genetic isolation. Genetic isolation may also result from *polyploidy*.

isotonic (of a solution) Not causing any change in the volume of an animal cell. Isotonic solutions usually have the same *water potential* as the cell.

isotope A variety of a chemical element that has the same atomic number as another variety of the same element and identical chemical properties, but differs from it in atomic weight. Radioactive isotopes are often used in biology to follow the path of a substance through an organism; their progress can be traced, but they behave biologically in the same way as non-radioactive isotopes. Carbon dioxide radioactively labelled with ^{14}C, for example, is used to study the products of photosynthesis.

J See *joule*.

jaw A structure of bone or *cartilage* in vertebrates, that forms the framework of the mouth and opens and closes it. Jaws usually bear teeth or horny plates to facilitate seizure or mastication of food.

jejunum Part of the small *intestine*, between the *duodenum* and *ileum*, which has large *villi* and is the main area for absorption of nutrients from the *gut*.

joint The point of contact between two or more parts of a solid skeleton, such as the bones of a vertebrate endoskeleton or

segments of an insect exoskeleton. Most joints are at least partly moveable and are called articulations, but some vertebrate
joints are immovable (e.g. those between the bones which form
the *cranium*). In humans, moveable joints include the ball-and-
socket shoulder joint, which allows movement in several
planes; the hinge joint in the elbow, which allows movement
mainly in one plane only; and the gliding joints between vertebrae, which allow only limited movement. See also *synovial
joint*.

joule (J) The SI unit of energy, heat, and work (see Appendix
4). One kilojoule (kJ) = 1000 joules. Although use of the joule is
recommended for all scientific purposes, the calorie is still
used, especially (as the kilocalorie or Calorie) in studies of diet
and nutrition (1 cal = 4.2 J; 1 Calorie = 1000 cal = 4.2 kJ).

J-shaped growth curve A *graph* curve that shows growth
increasing very rapidly against time then declining suddenly. It
is exhibited by populations of species (e.g. aphids) whose
growth rate is *density-independent* until the final decline,
which may be due to food exhaustion, seasonality of breeding,
emigration, or some other factor. Such species are called
J-species and their population growth is described as 'boom
and bust'. See also *r-species*.

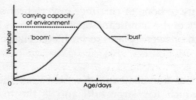

J-shaped growth curve

keratin A tough, fibrous *protein*, which consists of single
polypeptide chains each with an alpha-helix *secondary*

structure (this is like the shape of an extended spring). Disulphide bonds (very strong sulphur-containing bonds) strengthen the chains. Keratin is found in hair, teeth and nails. The disulphide bonds are less frequent in hair than in nails, making the former more flexible than the latter.

key A set of alternative statements from which an unknown organism can be identified on the basis of its visible features. See also *dichotomous key*.

kidney The main organ of nitrogenous excretion and *osmoregulation* in vertebrates. Mammals have a pair of kidneys at the back of the abdomen, made up of units (*nephrons*) each of which has a *glomerulus* that acts as an ultrafiltration unit. The filtrate which passes down the nephron is processed and eventually forms urine which is eliminated from the body. By varying the composition of the urine, the kidney carries out its main functions. These include excretion of the nitrogenous wastes (mainly *urea*), osmoregulation, ionic regulation (regulation of body salts), and maintenance of acid—base balance of the body.

kilocalorie See *calorie* and *joule*.

kilojoule One thousand *joules*.

kinesis An orientation behaviour in which the rate of movement of a cell or organism depends on the intensity of the stimulus, but not its direction (e.g. a woodlouse moves faster in a dry environment than in a damp one, thus tending to keep it in a moist habitat).

kinetic energy The energy of motion. It is energy actually in the process of doing work. Atoms and molecules move randomly; the faster they move, the greater their kinetic energy.

The total kinetic energy of the molecules of a particular substance is proportional to its concentration. This is why molecules diffuse from a higher concentration to a lower concentration (see *diffusion*).

kingdom The highest taxonomic group used in the classification of organisms. Many classifications are in use, all more or less subjective. The one used here is the five-kingdom classification proposed by L. Margulis and K.V. Schwartz (*Five Kingdoms: An illustrated guide to the phyla of life on Earth*, 1988). The five kingdoms are *Prokaryotae*, *Protoctista*, *Fungi*, *Plantae* and *Animalia*.

kinin See *plant growth substance*.

kneecap See *patella*.

Krebs cycle See *aerobic respiration*.

K-species A species that has a relatively low potential rate of reproduction influenced by population density (see *density-dependence*), keeping its numbers close to the *carrying capacity* of its environment. It has an S-shaped population growth curve. K-species tend to be large, long-lived organisms such as oak trees and humans, that live in stable *habitats*. Compare *r-species*.

labia See *vulva*.

lacteal One of the small *lymph* vessels in the centre of a *villus* in the small *intestine* that absorb neutral fats in the form of a milky white emulsion.

lactic acid A three-carbon organic acid with the formula $CH_3.CHOH.COOH$. It is produced from *pyruvate* as a by-product of *anaerobic respiration* in animals and some micro-organisms (lactic acid produced by lactose-feeding bacteria makes milk taste sour). In mammals, lactic acid can be converted in the *liver* to *glucose* or back into pyruvate when oxygen is available for *aerobic respiration*. Lactic acid is thought to contribute to the development of fatigue in muscles.

lactose or **milk sugar** A disaccharide formed by a *condensation* reaction that combines *galactose* and *glucose*. The *enzyme* lactase accelerates the *hydrolysis* of lactose into its components.

Lamarckism A theory of *evolution* proposed by the French naturalist Jean Lamarck (1744—1829), who suggested that characteristics an organism acquires during its lifetime can be passed on to the next generation (sometimes called the 'inheritance of acquired characteristics'). Expressed in modern terms, it would mean that changes in *phenotype* could determine the *genotype* of future generations. This does not agree with modern genetics and there are no clear examples of such inheritance. Consequently, the theory has fallen into disuse.

lamella (*pl.* **lamellae**) A thin, plate-like structure. The term is used, for example, of the sheet-like membranes between the grana in *chloroplasts*, the spore-bearing gills of mushrooms and toadstools, the plate-like structures in the gills of fish, and the thin, concentric rings that form compact bone.

lamina See *leaf*.

large intestine See *alimentary canal* and *intestine*.

larva (*pl.* **larvae**) The juvenile form of an animal, which is structurally and functionally distinct from the adult and often occupies a different habitat, allowing different resources to be exploited. Examples are tadpole and frog, caterpillar and butterfly. The larval stage may serve as a dispersal or feeding phase, as a resistant phase in which the organism survives adverse conditions, or as an asexual reproductive phase.

larynx A structure located at the junction between the *trachea* and *pharynx* in land-dwelling vertebrate animals. In mammals, it contains a pair of elastic membranes (vocal cords), which can be made to vibrate and so produce sound. Muscles in the larynx move *cartilage* within its walls, altering the tension of the vocal cords and varying the pitch of the sound produced.

latent heat of vaporization The heat required to change the state of a unit mass of a substance from a liquid to a gas, without a change of temperature. Water has a high latent heat of vaporization which makes it a very good coolant (e.g. in sweating).

law (in science) A rule or generalization about particular natural phenomena that is supported by a large body of experimental evidence. Although there are exceptions to many so-called laws, a true scientific law should have no exceptions.

law of independent assortment A law formulated by Gregor Mendel, stating that (in modern terms) each of a pair of alleles may combine randomly with either of another pair (i.e. the alleles of one *gene* segregate into the *gametes* independently of the way alleles of other genes have segregated). This is Mendel's second law, based on his studies of the inheritance of two pairs of characteristics (*dihybrid inheritance*).

law of segregation A law formulated by Gregor Mendel, stating that (in modern terms) the *characteristics* of a diploid organism are determined by alleles (Mendel's 'factors') occurring in pairs. Of a pair of such alleles, only one can be carried in a single *gamete*. The alleles normally pass unchanged from parent to offspring. This is Mendel's first law, based on his studies of the inheritance of single pairs of contrasting characteristics (*monohybrid inheritance*).

leaching The removal of substances from soil by the downward movement of water, by which they are carried into the drainage system of rivers and lakes. Leaching impoverishes soils of soluble nutrients. Nitrates, often added as artificial fertilizers to soils, are easily leached and can accumulate in aquatic habitats to cause *eutrophication*. Pollutants dumped on land can also leach into water courses and may even end up in drinking water.

leaf An outgrowth of a plant stem that is adapted for *photosynthesis* and is the region through which most water is lost by *transpiration*. Most leaves are flat, green structures with a large surface-area-to-volume ratio that maximizes light absorption. In flowering plants, the thin, broad, blade-like part of the leaf is called the *lamina*; it may be connected to the main plant stem by a petiole (leaf stalk) and grows from a lateral bud formed at the axil (the angle between the upper side of a leaf and the stem on which the leaf is growing). Vascular bundles extend into leaves as a series of veins that provide mechanical support and also transport materials into and out of the leaf. See diagram overleaf.

leaf litter Semi-decomposed plant material that forms a layer (the litter layer) covering the soil in areas of high leaf-fall (e.g. woodland). The layer may be up to 50 mm deep beneath trees whose leaves decompose slowly. The litter provides an important source of organic material for soil organisms and tends to maintain a high humidity at the soil surface. See diagram overleaf.

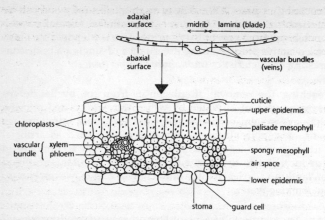

leaf Transverse section through a leaf blade.

learning *Behaviour* in which an organism consistently modifies its response to a stimulus or set of stimuli, usually to its own advantage, as a result of experience. Learning allows organisms to respond flexibly to situations. Categories of learning include associative learning through conditioning (see *reflex*), *habituation*, *imprinting* and *insight learning*.

legume 1. A dry fruit (e.g. of peas, beans or gorse), formed from a single *carpel*, in which the fruit or pod splits along two sides to release the seed.
2. A plant belonging to the family Leguminosae. Some (e.g. peas and beans) have *root nodules* containing nitrogen-fixing bacteria.

lens See *eye*.

lenticel (in plants) A small pore on the surface of woody stems and roots through which gas exchange takes place. Lenticels usually appear as raised spots surrounded by a powdery material which is probably *cork*.

leucocyte See *white blood cell*.

leukaemia Cancer of the tissues that form blood cells, in which there is usually an uncontrolled overproduction of one type of *white blood cell* and suppression of other types.

lichen An organism formed by the union of a green *alga* or cyanobacterium (blue-green alga) and a fungus which live mutualistically (see *mutualism*). The alga or cyanobacterium produces oxygen and carbohydrates for the fungus; the fungus can conserve water and provides water, carbon dioxide, and mineral salts for its partner. Lichens are slow-growing and have a simple, undifferentiated structure, which can be scaly, crust-like, leafy or shrubby depending on the species. Lichens are found encrusting trees, rocks and buildings. They often occur in very inhospitable environments. Lichens can be used as pollution monitors: some species are very sensitive to air pollution, especially sulphur dioxide gas (a common waste product of industrial processes), and the abundance and variety of lichens increase as pollution decreases.

life The state of physical beings (called organisms) that use substances obtained from outside themselves (see *nutrition*) in order to grow, maintain their bodies, and reproduce, and that exhibit *excretion*, *respiration*, movement and *irritability*.

life cycle The series of stages through which individuals of a population pass, including fertilization, reproduction, and the death of those individuals and their replacement by a new generation.

ligament *Connective tissue*, containing *collagen*, that binds two bones together.

ligase An *enzyme* that catalyses the joining together of two molecules. Such reactions require energy, usually obtained from the simultaneous breakdown of *ATP*. A special ligase (DNA ligase) is used in *genetic engineering* to join together pieces of DNA.

light Electromagnetic radiation from the Sun. The visible spectrum of light (the electromagnetic spectrum) ranges from red light (wavelength about 700 nm) to violet light (wavelength about 400 nm). Sunlight is harnessed by plants for *photosynthesis* and forms the primary energy source for most organisms on the Earth.

light-dependent stage The stage of *photosynthesis* that can take place only in the presence of light. In plants, it occurs on the grana of *chloroplasts*, where *chlorophyll* absorbs light energy. An electron in the chlorophyll is excited, escapes, and is used for either cyclic *photophosphorylation* or non-cyclic photophosphorylation. During cyclic photophosphorylation, the electron is passed along a series of *electron carriers* and, after some of its energy is used to make *ATP*, returns to the chlorophyll. In non-cyclic photophosphorylation, the electron is passed along a series of electron carriers to make ATP, but then is taken up by *NADP*. This enables water to be split into hydrogen ions (protons) and hydroxyl ions (a process known as photolysis). The hydrogen combines with NADP to form NADPH. The hydroxyl ions combine to form oxygen and water, and the hydroxyl ions also donate electrons to restore the stability of chlorophyll. Hydrogen from NADPH, manufactured during the light-dependent stage, is used to reduce carbon dioxide during the *light-independent stage*. ATP provides the energy required for this reduction which results in the synthesis of carbohydrates.

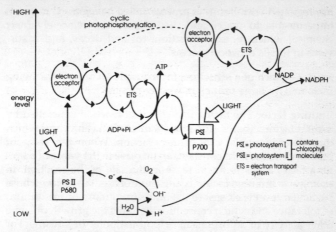

light-dependent stage The main chemical pathways in the light-dependent stage of photosynthesis.

light-independent stage or **dark stage** The stage of *photosynthesis* which can take place in the absence of light, provided there is a sufficient supply of NADPH and *ATP*. In plants, it takes place in the *stroma* of *chloroplasts* and results in the reduction of carbon dioxide to make carbohydrates during the *Calvin cycle*.

lignification See *lignin*.

lignin A complex polymer found in many plant *cell walls*. It is neither a *carbohydrate* nor *protein*, but is derived from the oxidation of tyrosine, an *amino acid*. Its function appears to be to cement together and anchor *cellulose* fibres and to stiffen cell walls. Lignin adds considerably to the mechanical strength of plant cells. It forms 20—30% of the wood of trees. Lignin also lines the walls of *xylem* vessels; when they become completely impregnated with lignin (a process called *lignification*) they are impermeable to water and die. Lignin is a relatively inert chemical which reduces infection, rot and decay, and it survives in fossils of woody stems.

limb A branch of a structure, for example an arm, leg or wing in animals; a large branch in woody plants.

limiting factor The factor, in a chemical reaction controlled by several factors, that is closest to its minimum value, and which thus determines the rate of the reaction. When the value of other factors are held constant, an increase in the value of a limiting factor leads to an increase in the rate of the reaction. In ecology, a limiting factor is the environmental factor whose particular level is closest to the limits (minimum or maximum) of tolerance and which consequently limits the growth or some other activity of an organism or population of organisms. For example, the growth of crops may be limited by nitrate deficiency when other factors would favour a high growth rate.

limits of tolerance The upper and lower limits to the range of particular environmental factors (e.g. pH, temperature, light, salinity, water availability) within which an organism can survive.

line graph A diagrammatic representation of the relationship between two or more variables in which only pure numbers are plotted. See appendix 1.

line transect See *transect*.

linkage A measure of the degree to which different *genes* are inherited together. Genes occurring on the same *chromosome* are said to belong to the same linkage group. Their linkage usually depends on how far apart they are: genes close together tend to remain together during the separation of *homologous chromosomes* at *meiosis*.

linkage group See *linkage*.

lipase An *enzyme* that accelerates the breakdown of *lipids* into *fatty acids* and *glycerol*. In mammals, lipase is secreted by the *pancreas* and digests lipids in the *duodenum*. Lipases are also secreted in adipose cells where they break down triglycerides into glycerol and fatty acids, enabling fat stored in the cells to be mobilized as an energy source.

lipid A member of a very variable group of organic substances. Lipids are defined loosely as organic substances which do not dissolve in water, but which can be extracted from cells by organic solvents such as ether, chloroform, and benzene. True lipids are esters, i.e. organic compounds formed by a reaction between acids and an alcohol, e.g.

$$CH_3COOH + C_2H_5OH \rightarrow CH_3COOC_2H_5 + H_2O$$

    fatty acid   alcohol   ester   water

The process, called esterification, takes place by a condensation reaction (i.e. a reaction which results in the formation of water). See also *fatty acid*.

liquid feeder An animal that feeds exclusively on fluids or dissolved substances. Examples are aphids that suck sap from *phloem*; humming birds, worker bees, and butterflies that drink nectar; mosquitoes that suck blood; and houseflies and spiders that liquidize their food before ingestion.

litter layer See *leaf litter*.

liver A large organ in vertebrates occupying much of the upper part of the abdomen, which receives all the blood from the small *intestine* and processes materials absorbed from that part of the gut. It has many important homeostatic functions (see *homeostasis*), including *deamination* of blood to form *urea* from nitrogenous wastes; detoxification of blood; storage of sugars and some vitamins and minerals (e.g. in humans, vitamins A, D, and B12, iron, copper, and potassium); manufacture of plasma proteins (e.g. *fibrinogen*), *cholesterol*, and *bile*; and, in *endotherms*, production of heat.

loam A rich, friable, *humus*-rich soil consisting of sand, silt, and clay in roughly equal proportions. It is an easily worked soil much sought after by farmers.

lock-and-key theory A theory of *enzyme* action which proposes that a *substrate* fits into the *active site* of an enzyme like a key into a lock. The shape of the substrate ('key') must fit that of the active site ('lock') exactly. It is thought that the active site has some regions which hold the substrate in place and other regions containing atoms that speed up the chemical reaction. The theory helps explain why enzymes are specific and why a change in shape of either the substrate or enzyme affects the enzyme's action, but it does not provide a totally satisfactory explanation of enzyme action. For example, in the lock-and-key theory, enzyme action depends on the unlikely event of substrate molecules making very precise contact with the active site. The induced-fit model of enzyme action, a modified version of the lock-and-key theory, does not depend on such precise contact being made; the active site is able to change its shape. When substrate molecules approach an enzyme, the shape of the active site changes so it can enfold the substrate and assume its most effective catalytic shape.

lockjaw See *tetanus* (2).

locomotion The movement of an organism from one place to another independently of any outside force. Most animals do this to acquire food, but most plants are immobile because they make their own food using the raw materials in their immediate surroundings and the energy from the Sun.

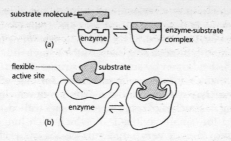

lock-and-key theory (a) lock-and-key model of enzyme action;
(b) induced-fit model of enzyme action.

locus The position of a *gene* within a *DNA* molecule, or the specific position a particular gene occupies on a *chromosome*.

logarithmic scale A scale in which the values of a variable are expressed as logarithms. Logarithmic scales are often used in *graphs* where one variable changes exponentially (e.g. in population growth).

longitudinal section A section cut lengthwise through a specimen, from anterior to posterior (front to rear) or from top to bottom.

long sightedness or **hypermetropia** An *eye* defect in humans, usually caused by the distance between the lens and retina being too short, so the focal point of light is behind the retina. It is corrected using convex lenses to converge the light rays and focus them on the retina.

loop of Henle See *nephron*.

lumen A cavity enclosed within a structure or cell. Food passes through the gut lumen. See *peristalsis*.

lung The organ of *gas exchange* in amphibians, reptiles, birds and mammals. In mammals, the lungs are a pair of elastic sacs in the chest cavity, linked to the air by a system of tubes. Air enters the mouth or nose and passes into a single tube (the trachea) that is held open by semi-circular rings of *cartilage*. The trachea branches into two *bronchi*, one serving each lung. Bronchi branch into *bronchioles*, the diameter of which can vary, terminating in bunches of minute, round sacs (called *alveoli*; the Latin word for 'small cavities'), of which each lung contains millions, providing a very large surface for gas exchange by *diffusion*. Each alveolus is only one cell thick and covered in small capillaries, further facilitating diffusion of the respiratory gases. The lungs are inflated and deflated by movements of the ribs (see *intercostals*) and *diaphragm*. Inhalation draws fresh, oxygen-rich air into the lungs; exhalation carries carbon-dioxide-rich air out of the lungs.

luteinizing hormone A *hormone* secreted by the anterior *pituitary* gland in mammals. In mature males, it stimulates *testosterone* production in the testes. In mature females, it stimulates oocyte development, *ovulation*, and the *corpus luteum* (a temporary *endocrine gland* formed from the *Graafian follicle* after ovulation).

lymph A milky or colourless fluid which drains from the *tissue fluid* into the *lymphatic system*. It is similar to the tissue fluid in composition, but tends to contain more fats and large numbers of *white blood cells*.

lymphatic system In vertebrates, a system of blind-ending tubules that drains excess tissue fluid (*lymph*) and transports it to two veins in the upper thoracic cavity. Except in birds and mammals, *lymph hearts* pump lymph through the system. In birds and mammals, lymph flow is achieved by the action of adjacent skeletal muscles and is maintained in one direction by one-way valves. The lymphatic system of mammals contains clumps of tissue (*lymph nodes*) that act as filters and are sites of *white blood cell* formation.

lymph heart See *lymphatic system*.

lymph node See *lymphatic system*.

lymphocyte A *white blood cell* which has a large nucleus and a clear *cytoplasm*. In mammals, lymphocytes are produced in the bone marrow and lymph tissue, and are present in large numbers in the *lymphatic system* (especially at lymph nodes) and blood. Lymphocytes play an important role in the formation of protective substances (see *antibody*) which help defend the body against invasion by foreign substances.

lysis 1. The breakdown or dissolution of cells.
2. (-lysis) A suffix denoting breakdown (e.g. hydrolysis and photolysis).

lysosome A cell *organelle* consisting of a simple spherical sac bounded by a single membrane and containing hydrolytic *enzymes*. Lysosomes are involved in the breakdown of structures or molecules. They destroy unwanted organelles (e.g. mitochondria), help digest material taken in by *endocytosis*, and they sometimes release their contents by *exocytosis* to digest extracellular material (e.g. during bone remodelling). Lysosomes are sometimes called 'suicide bags', because they can also release their enzymes inside the cell and digest the latter as a whole. This usually takes place after a cell dies and may have some adaptive value as a built-in mechanism for the self-removal of dead cells in a multicellular organism.
Lysosomes are thought to be produced by the *Golgi apparatus* (which makes the membrane) and the rough endoplasmic reticulum (which makes the hydrolytic enzymes).

macerate To soften or wear away tissue by *hydrolysis* or some other means.

macroevolution See *evolution*.

macronutrient A **nutrient** required by an organism in large amounts. In the human diet, macronutrients include carbohydrates, fats and proteins. Plant macronutrients include potassium, calcium, nitrogen, phosphorus, magnesium, sulphur and iron. See **mineral**.

magnesium An element that is essential for plants and animals. Magnesium acts as a *cofactor* for many *enzymes*, especially those catalysing the transfer of phosphate groups during *respiration*. In mammals, it plays a role in nerve conduction and muscle contraction. In plants, it forms part of the *chlorophyll* molecule and can release electrons during the light-dependent stage of photosynthesis. Deficiency causes *chlorosis* in green plants (a yellowing of leaves due to lack of chlorophyll).

maintenance The act of keeping a biological system stable: the system may be a cell, an organism or part of one, an *ecosystem*, or the *biosphere*. It requires an input of raw materials and energy from the environment and involves homeostatic mechanisms, usually incorporating negative *feedback*. Maintenance is a fundamental requirement of life; organisms must maintain themselves for long enough to reproduce or they will become extinct.

malleus or **hammer** A hammer-shaped bone that forms one of the *ear ossicles*.

malnutrition Poor nutrition caused by an insufficient or inappropriate diet. It may result from nutrients being deficient, in excess, or not in the correct proportions.

Malpighian layer A layer in the epidermis of vertebrate skin where cell division is active and *melanin* (a black pigment) may occur in cells.

maltase An *enzyme* that catalyses the breakdown of *maltose* to two *glucose* molecules.

maltose or **malt sugar** A disaccharide *reducing sugar* formed by the combination of two *glucose* molecules linked together by a *glycosidic bond*. Maltose is also a product of digestion of *starch* and *glycogen* by *amylases*.

malt sugar See *maltose*.

Mammalia or **mammals** A class of endothermic, chordate animals (see *endotherm*, *Chordata*), which have skin with hair follicles and sweat glands. They are viviparous (the mother produces active young rather than eggs) and the young are fed on milk.

mammals See **Mammalia**.

mammary gland A gland in female mammals that produces milk used to suckle the young. A mammary gland usually has a conical projection, either a *nipple* with many openings or a teat with a single opening.

mandible 1. (in vertebrates) The lower *jaw*.
2. (in some *arthropods*) One of a pair of feeding mouthparts.

Mann—Whitney U test A statistical test to determine the significance of differences between population medians of two samples of numerical data put into rank order. The data of the two samples should be unmatched (i.e. they cannot be arranged in matched pairs) and the size of each sample should be 8—20 inclusive.

mark—release—recapture A technique for estimating the population density of more or less mobile animals in a clearly defined area (e.g. a pond). A sample of the population is captured, marked, and released, and the proportion of marked individuals (recaptures) in a subsequent sample is recorded. The population density can be estimated using the Lincoln—Peterson method: $N = S_1 S_2 / R$, where N is the estimated population size (known as the Lincoln Index), S_1 is the number of individuals caught and marked in the first sample, S_2 is the total number of individuals in the second sample, and R is number of marked individuals in the second sample (recaptures). This estimate is reasonably accurate only if the following conditions are met: (a) at least 10% of the marked and released individuals are in the second sample; (b) marked individuals become completely mixed in the population; (c) the marking does not harm the animal; (d) no animals lose their mark; (e) marked individuals have the same chance of being captured as unmarked ones; (f) no emigration or immigration takes place; and (g) there are no births or deaths among the two samples.

mass flow The transport of substances by a fluid which is moving from one location to another along a hydrostatic pressure gradient. Streams carry their load of sediment and dissolved substances by mass flow, and blood plasma, pumped out of the heart at high pressure, transports blood components to the rest of the body, where pressure is lower. Mass flow has also been proposed as an explanation for *translocation* in plants; it is suggested that *solutes* are carried in the flow of water in *phloem* from regions of high hydrostatic pressure, where actively photosynthesizing cells cause an osmotic influx of water, to regions of low hydrostatic pressure where the storage or use of photosynthetic products results in water being lost.

mastication The mechanical breakdown of food by chewing and grinding, prior to swallowing. Mastication is very important in herbivorous mammals, which have large premolars and *molars* adapted for the process.

mating 1. Sexual reproduction that involves two individuals.
2. (in mammals, birds, and some other animals) The behavioural process of pair formation which leads to sexual reproduction. See also *copulation*.

matrix 1. (in *connective tissue*) The non-living substance (ground substance) between cells.
2. (in *mitochondria*) The inner, fluid region, enclosed within the inner membrane, containing enzymes, DNA, and ribosomes.
3. The medium in which a substance is embedded.

maturity The final stage in the development of a cell, organism, or ecosystem.

maxilla 1. (in vertebrates) The upper *jaw*.
2. (in some *arthropods*) An appendage close to the mouth that is modified for feeding.

mean or **arithmetic mean** A measure of the average value of the data comprising a sample. The mean of a sample is the sum of the measurements divided by the number of measurements (sample size). See also *median* and *mode*.

median The central value in a sample; half of the sample lies above it and half below it. See also *mean* and *mode*.

medulla oblongata See *brain*.

megaspore See *heterospory*.

meiosis or **reduction division** Nuclear division which results in the production of daughter nuclei, each containing half the number of sets of *chromosomes* of the parent nucleus. Meiosis is the basis of sexual reproduction. In mammals, it results in the production of haploid *gametes*. In most plants, it results in the production of haploid *spores*. Typically, a parent cell with two chromosome sets divides to form four daughter nuclei each with one chromosome set. Unlike those produced in *mitosis*, the daughter nuclei are not identical with the parent nucleus. Meiosis involves two successive divisions, each of which, for descriptive purposes, is divided into four stages. These occur in the following order: prophase I (*homologous chromosomes* pair and *crossing-over* occurs); metaphase I (chromosomes line up in homologous pairs on the equator); anaphase I (homologous chromosomes separate); and telophase I (two daughter nuclei are formed), comprise the first division; and prophase II, metaphase II, anaphase II, and telophase II comprise the second division. The first division (meiosis I) leads to the separation of homologous chromosomes and reduction in the number of chromosome sets; the second division (meiosis II) is concerned with the separation of sister *chromatids*. Crossing-over during prophase I results in the exchange of alleles between each pair of homologous chromosomes and increases the genetic variability of daughter nuclei.

melanin See *melanism*, *Malpighian layer*.

melanism Excessive development of the black pigment, *melanin*, in animals. Industrial melanism is the increased frequency of darkly pigmented animals (e.g. the peppered moth, *Biston betularia*) due to directional selection (see *natural selection*) in areas blackened with soot.

melting point The temperature at which a solid begins to change into a liquid. See also *water*.

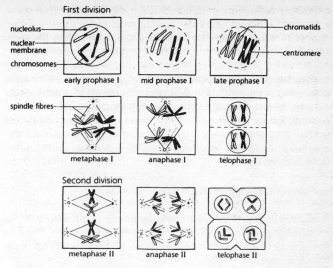

meiosis The stages of meiosis in a cell containing two pairs of homologous chromosomes.

membrane A thin sheet, skin, or layer of tissue covering a part of an animal or plant. See also *cell membrane*.

memory Information related to previous experience that is stored in an organism or cell and can be retrieved and used in future situations, or the ability to store and retrieve such information. Memory is an essential component of *learning*.

Mendelian inheritance Inheritance which conforms to Mendel's first and second laws (see *law of segregation, law of independent assortment*). Typically, Mendelian inheritance results in offspring with simple *phenotype* ratios (e.g. 3:1 in *monohybrid inheritance,* and 9:3:3:1 in *dihybrid inheritance*).

meninges Membranes enclosing the *brain* and *spinal cord*.

menopause In human females, the time in life when the *menstrual cycle* and ovulation become irregular and then cease.

menstrual cycle The approximately monthly cycle of changes associated with ovulation that occur in the *uterus* of sexually mature, non-pregnant human females and some other closely related female mammals. It is characterized by ovulation followed by the breakdown of the wall of the uterus and discharge of blood and fragments resulting from this breakdown (*menstruation*).

menstruation See *menstrual cycle*.

meristem The part of an actively growing plant where nuclei are dividing by *mitosis* and new permanent plant tissue is forming. Apical meristems occur at the tips of shoots and roots; lateral meristems include *cambium*.

mesocarp See *fruit*.

mesoderm The middle layer of cells in an *embryo*, lying between the *ectoderm* and *endoderm*, which gives rise to the muscles in flatworms and lines the *coelom* in coelomates. In *vertebrates*, it gives rise to muscles, blood vessels, much of the kidneys, and the dermis of the skin.

mesoglea See *Cnidaria*.

mesophyll The middle layer of cells in a leaf, lying between the epidermal layers. Mesophyll cells carry out *photosynthesis*. There are two main types: *spongy mesophyll* and *palisade mesophyll*. Spongy mesophyll consists of spherical cells, containing few *chloroplasts*, which are loosely packed with large air spaces connected to the atmosphere through the stomata. The palisade layer usually lies immediately below the upper *epidermis*. It consists of densely packed cells, elongated at right angles to the leaf surface, containing numerous chloroplasts, and its main function is photosynthesis.

mesophyte A plant that is adapted to live in soils of average mineral salt content which are neither extremely wet nor extremely dry. Most terrestrial flowering plants are mesophytes. They tend to wilt during droughts, because they have no special adaptations to conserve water. Compare *halophyte*, *hydrophyte* and *xerophyte*.

mesosome An *invagination* (infolding) of a bacterial cell surface membrane on which enzymes associated with aerobic respiration are found.

messenger RNA See *RNA*.

metabolic pathway A series of interconnected biochemical reactions in living cells, in which the product of one reaction is the raw material for the next. Metabolic pathways may be linear (e.g. *glycolysis*) or cyclical (e.g. *Calvin cycle*). The chemical changes and energy transformations in each reaction are usually small and catalysed by specific **enzymes**. The output of end-product from the pathways is often controlled by negative *feedback*.

metabolic rate See *metabolism*.

metabolic waste Any useless by-product of *metabolism*. Some waste products (e.g. *urea* in mammals) may be harmful if they accumulate in the body and must be eliminated (excreted).

metabolism The sum of all the chemical reactions, including *anabolic reactions* and *catabolic reactions*, which take place in an organism and sustain its life. The rate at which metabolism takes place (*metabolic rate*), varies according to the species of organism, its sex, age, size and activity, and is usually expressed as energy per unit surface area per unit time (e.g. $kJm^{-2}h^{-1}$) and can be estimated by *calorimetry*. The basal metabolic rate (BMR) refers to the minimum amount of energy that can sustain a non-digesting organism at rest in an environment at the same temperature as its own body.

metameric segmentation The subdivision of the body of an animal into a series of discrete units (segments) which are essentially similar to one another. Unlike the segments in, for example, a tapeworm, metameric segments are fixed in number and are all of the same age. Metameric segmentation is most clearly displayed by annelid worms (see *Annelida*). Each worm segment usually has a similar pattern of blood vessels, nerves and musculature, but the pattern may be modified by specialization of some body regions. Metameric segmentation is thought to have evolved as a means of improving efficiency of locomotion; the division of the body musculature and nerves into separate units would have aided the passage of rhythmic waves of muscle contractions along the body.

metamorphosis The process by which some animals (e.g. butterflies and frogs) undergo a dramatic physical change from the larval to the adult condition.

metaphase A stage in nuclear division when *chromosomes* line up at the equator of the cell. In *meiosis* I, *homologous chromosomes* are arranged in pairs along the equator. In *mitosis* and meiosis II, chromosomes are arranged singly along the equator and the *centromeres* divide before the sister *chromatids* are separated.

microbe or **microorganism** An organism that can be seen individually only with the aid of a microscope. See *bacteria*, *unicellar organisms* and *virus*.

microbiology The scientific study of the structure and function of *microbes*.

microclimate See *habitat*.

microevolution See *evolution*.

microhabitat See *habitat*.

micronutrient A *nutrient* required by an organism only in very small or trace amounts. Micronutrients in the human diet include *vitamins* and *trace elements*. Plant micronutrients include the *minerals* boron, copper, magnesium, molybdenum and zinc.

microorganism See *microbe*.

micropyle A canal, in flowering plants, formed by the *integuments* of an ovule, through which the *pollen tube* enters the *embryosac*. In a seed, the micropyle forms a minute pore through which water enters at the start of *germination*.

microscope An instrument for producing magnified images of an object. Optical microscopes (or light microscopes) use light radiation to form an image. The simplest type is a hand lens, but the most powerful types are compound microscopes consisting of two sets of lenses: eyepiece lenses and objective lenses. By using mirrors or an internal light source, light is either reflected from the surface of an object or transmitted through it so that an image can form directly on the retina. The maximum effective magnifying power of a normal light microscope is about 1500 x. Its power is limited by the wavelength of light. Higher magnifications can be obtained with the *electron microscope* which uses an electron beam to produce high resolution images of a specimen (see *resolving power*). Electron microscopes can produce clear, well-defined images of greater than 500 000 x magnification. A typical electron microscope consists of a tube containing a vacuum in which a beam of electrons are 'fired' from an electron gun, consisting of a hot filament. The electrons are guided onto or through the specimen by powerful electromagnets. There are two main types of electron microscope. The *transmission electron microscope* (TEM) passes a beam of electrons through a thinly sectioned specimen to produce an image on a fluorescent screen, or onto a photographic plate to form a black and white photograph (an *electron micrograph*). The *scanning electron microscope* (SEM) forms a three-dimensional image from electrons reflected from the surface of a specimen which may be whole cells, tissues or organisms.

microspore See *heterospory*.

microtome See *sectioning*.

microtubule A very fine, hollow, cylindrical tube in the *cytoplasm* of a cell, made up of small units of a *globular protein* called tubulin. Microtubules are a component of the *cytoskeleton*, *spindle apparatus*, *cilia* and flagella (see *flagellum*). In addition to helping to maintain the shape of the cytoplasm, microtubules are thought to be involved in movement of materials and *organelles* within cells.

microvillus (*pl.* **microvilli**) One of a number of minute, finger-like extensions of the cell surface membrane of some animal cells (e.g. cells lining the small intestine). Microvilli increase the surface area of the cell for absorption. Each microvillus contains contractile proteins (*actin* and *mysosin*) that enable it to move.

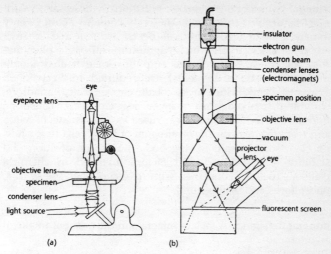

microscope Comparison between an optical microscope (left) and an electron microscope (right).

middle ear See *ear*.

milk A whitish fluid, secreted by the *mammary glands* of mammals, which provides a balanced *diet* for young offspring. See also *breast-feeding*.

milk sugar See *lactose*.

milk teeth or **deciduous teeth** The first of two sets of teeth in mammals, which are shed in early life and in some species before birth. The milk teeth are smaller and fewer in number than the second, permanent set of teeth.

mineral An inorganic element, nutritionally necessary, which must be derived from food. The body contains about twenty minerals. Eight (called *macronutrients*) are required in large amounts: calcium, chlorine, magnesium, nitrogen, phosphorus, potassium, sodium and sulphur. Some (called micronutrients or trace elements) are required in very small amounts. They include iron, cobalt, copper, zinc, vanadium, chromium, molybdenum, manganese, silicon, tin, selenium, fluorine and iodine. Minerals are used as constituents of bones and teeth (e.g. calcium); constituents of body cells (e.g. phosphorus); soluble salts giving body fluids composition and stability (e.g. sodium); and factors involved in chemical reactions (usually concerned with release of energy, e.g. iron, phosphorus and magnesium).

mineral deficiency A lack of minerals in the nutrient intake of an organism.

miscarriage The expulsion from a womb of a foetus that is not yet capable of independent survival.

mitochondrion (*pl.* **mitochondria**) An *organelle*, about 0.1—0.5 µm long, found in all *eukaryotes*, that is the site of *aerobic respiration*. It consists of two membranes enclosing a gel-like matrix containing *ribosomes*, phosphate granules, DNA, and *enzymes*, and is the site of the Krebs cycle. The inner membrane is folded inwards to form *cristae*, the walls of which are the sites of *electron transport* and *oxidative phosphorylation*. Staining reveals the presence of elementary particles on the inner membrane, each consisting of a head piece, stalk, and base. The head contains ATPase, the enzyme required for *ATP* synthesis. Mitochondria are especially abundant in highly active cells (e.g. those of the liver and muscles). They are self-replicating.

mitosis Nuclear division that results in the formation of two daughter nuclei which are identical to the parent nucleus. Mitosis forms the basis of *a sexual reproduction* and of growth and repair in multicellular organisms. Mitosis is preceded by interphase when the cell is prepared for nuclear division (e.g. DNA is replicated). Mitosis is a continuous process but, for the purposes of description, is usually divided into four stages: prophase, metaphase, anaphase and telophase. Compare *meiosis*.

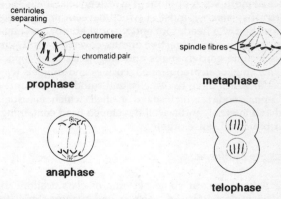

mitosis Stages of mitosis in a cell with two pairs of homologous chromosomes.

mode The most frequently occurring value among the data in a sample; it is a measure of the average. See also *mean* and *median*.

model A mathematical, physical, pictorial or computer representation of a system, used as an aid to understanding the system. If sufficiently accurate, the model can be used to predict the effects on the behaviour of the system of changes in selected factors. The accuracy and usefulness of a model are only as good as the assumptions on which it is based.

Monocotyledonae, monocotyledons or **monocots** A class of *Angiospermatophyta* (flowering plants) in which the *embryos* have a single seed leaf (cotyledon). Their mature leaves are usually elongated, lance-like, and have parallel veins. Monocots include cereals, grasses, lilies and palms. Compare *Dicotyledonae*.

monocotyledons See *Monocotyledonae*.

monoculture The cultivation over a large area of single plant species or (for crops) a single variety of plant. Monocultures have the advantage of producing plants of uniform characteristics and state of development, making management and harvesting more efficient. On the other hand, monocultures are generally more susceptible to pests and diseases.

monocyte A large *white blood cell* with a single, oval or horse-shoe-shaped nucleus that engulfs foreign bodies (e.g. bacteria) by *phagocytosis*.

monohybrid inheritance The inheritance of a single pair of contrasting *characteristics*. See also *law of segregation*.

monomer One of the chemical subunits which serve as building blocks for *polymers*.

monosaccharide or **simple sugar** A crystalline, sweet-tasting, very soluble *carbohydrate* which consists of a single chain or a single ring structure. Monosaccharides have the general formula $(CH_2O)_n$, where n = 3 or a larger number. Examples include glyceraldehyde (triose, n = 3), ribose (pentose, n = 5), and glucose (hexose, n = 6). Monosaccharides are very rich sources of energy and are the main fuels for cellular work.

morphology The study of the external structure, shape and form of organisms, as opposed to that of their functions (physiology) or internal structure (anatomy).

mortality Death rate, expressed as the percentage or numbers per thousand of organisms within a population that die each year.

motile Able to move from one place to another independently.

motor neurone See *neurone*.

mouth See *alimentary canal*.

mucin See *mucus*.

mucous membrane Tissue containing *goblet cells* that secrete *mucus*.

mucus 1. Any slimy or sticky material secreted by an organism. **2.** A slimy solution secreted by *goblet cells* in vertebrates, for protection and lubrication. It contains mucin, a *protein* combined with a *polysaccharide*.

multicellularity The arrangement of the body of an organism into more than one cell, so there is a division of labour and the organism can increase in size. Division of labour means that particular cells specialize to carry out different functions of benefit to the whole organism. This can increase the efficiency of an organism and enable it to exploit new environments, but a specialized cell may lose its ability to carry out one or more basic physiological processes. The cells of a multicellular organism are therefore to some extent interdependent and the survival of the organism depends on good cooperation and communication between its cells.

multiple alleles See *gene*.

murein See *bacteria*.

muscle 1. Fleshy, contractile tissue that moves parts of an animal body. There are three main types of muscle: *cardiac muscle*, *smooth muscle*, and striated muscle (see *skeletal muscle*). Muscle cells contain the contractile proteins *myelin* and *actin*. **2.** An animal structure comprising numerous striated muscle cells wrapped in *connective tissue* and supplied with blood vessels and nerves. There are approximately 600 muscles in the human body.

mutagen A chemical substance (e.g. colchicine or mustard gas) or other agent (e.g. ionizing radiation) that causes a change in the structure of *DNA* or chromosomal damage. See also *mutation*.

mutant An organism, cell, chromosome, or DNA that has undergone a *mutation*.

mutation A sudden, random change in the genetic material of an organism that may be inherited if it occurs in cells that produce the *gametes*. A mutation can occur as invisible changes in *DNA* (gene mutations), or as changes (visible under the light microscope) in the structure or number of *chromosomes* (chromosome mutations). Gene mutations occur when DNA is not replicated precisely or when a mutagen damages DNA; they may result from deletion (loss of a segment of DNA), insertion (insertion of an extra nucleotide base in an existing nucleotide sequence), substitution (replacement of one nucleotide base with another) or inversion (a reversal in the sequence of nucleotide bases in a DNA strand). Chromosomal mutations may result from deletions (loss of a segment of a chromosome), inversion (reversal of the gene sequence in a part of a chromosome), duplication (duplication of a chromosome segment on the same or another chromosome), aneuploidy (decrease in number of chromosomes within a set) or euploidy (increase in the number of sets of chromsomes; see also *polyploidy*). Most mutations are harmful, but a few increase the fitness of individuals and spread through a population by natural selection. Mutations are the ultimate source of genetic variation and are essential for *evolution*.

mutualism Any relationship between two individuals of different species that benefits both. An example is that between mammalian herbivores and gut bacteria, in which the mammal benefits from the cellulose-digesting ability of the bacteria and the bacteria benefit from the warm, protected environment and food supplied by the mammal.

mycelium A mass of *hyphae* which comprises the bulk of many fungi (e.g. mushrooms).

mycorrhiza An association between a fungus and the roots of certain plants; it is a form of *mutualism*. In some associations, the fungus forms coils inside the *cortex* of the roots and helps the plant absorb inorganic nutrients (e.g. phosphorus). In other associations, the fungus forms a sheath around the root and plays an important part in transferring organic substances to the plant. In both cases, substances synthesized by the plant are absorbed by the fungus. Some plants (e.g. certain orchids and conifers) appear unable to develop normally without a fungal partner.

myelin A fatty material that forms an insulating sheath around some vertebrate *neurones*, enabling nerve impulses to be conducted quickly. Myelin is formed from the membrane of *Schwann cells*, which is wrapped around a neurone in concentric layers.

myogenic (of contractions, especially of *cardiac muscle*) Arising spontaneously from within muscle cells and independently of nervous stimulation.

myoglobin A *protein* consisting of one polypeptide chain linked to an iron-containing *prosthetic group*. Myoglobin can combine reversibly with oxygen. It occurs in muscle cells where its oxygenated form (oxymyoglobin) acts as a store of oxygen for use when the oxygen released from *haemoglobin* is exhausted. Myoglobin gives muscle its red colour.

myopia or **short-sightedness** An *eye* defect in which light entering the eye is focused in front of the retina, so distant objects cannot be seen clearly. It is usually caused by the eyeball being too long and is corrected by using diverging (concave) lenses to focus light on the retina.

myosin A contractile *protein* found in the majority of cells of *eukaryotes*. In muscle cells, it combines with *actin* during contraction.

NAD *(nicotinamide adenine dinucleotide)* A coenzyme (see *cofactor*) that plays an important role as a *hydrogen acceptor* in *glycolysis* and the respiratory chain; it readily accepts and gives up hydrogen. In glycolysis, its oxidation and reduction is coupled to the production of *ATP*. NAD is derived from the *vitamin niacin* (one of the vitamins of the B complex).

NADP *(nicotinamide adenine dinucleotide phosphate)* A *hydrogen acceptor* and *electron carrier* that plays an important role in photosynthesis. NADP accepts the electron released from *chlorophyll* and the hydrogen released from water during the light-dependent stage of photosynthesis. These are then donated to the products of carbon dioxide to reduce them to sugars in the *Calvin cycle*.

nastic movement or **nasty movement** A growth movement of plants that occurs as a result of an external stimulus but does not take place in a particular direction. Daisy flowers, for example, open during the day in the presence of light and close at night when light is absent.

nasty movement See *nastic movement*.

natality Birth rate; the number of young produced per female per unit time (usually one year). In humans, natality is usually expressed as the number of births per thousand of population per year.

natural history The study of organisms in their natural environment.

natural selection The action of natural processes (e.g. predation, disease, competition for food, and temperature fluctuations) which result in the less well-adapted members of a *species* producing fewer offspring than better-adapted members. A predator, for example, may preferentially kill weaker prey, leaving only the stronger ones to breed and perpetuate their genes. Natural selection is the main driving force behind *evolution*, according to *Darwinism*. Natural selection acts on the observable variations of an individual (called the *phenotype*), such as height and colour. The frequency of each phenotype usually shows a *normal distribution*, described by a bell-shaped graph. There are three main types of natural selection. (a) Stabilizing selection tends to select against phenotypes at both extremes of the distribution. This occurs in selection for birth weight in humans: not surprisingly, abnormally small or large babies have low rates of survival under natural conditions; the highest survival rate is for babies around 3.4 kg, which is very close to the average birth weight. (b) Directional selection favours one extreme of the phenotype range. The long neck in the giraffe is believed to have evolved in this way. (c) Disruptive selection selects against intermediate phenotypes and favours those at the extremes, resulting in a bimodal distribution with two distinct groups of phenotypes. If the two groups are unable to interbreed, then each population may eventually give rise to a new species.

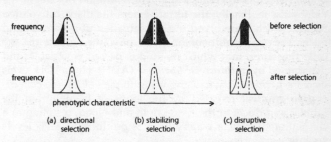

natural selection The three main types of natural selection acting on continuous phenotypic variations.

nectar See *flower*.

nectary See *flower*.

negative feedback See *feedback*.

nematoblast or **cnidoblast** A stinging thread cell. See *Cnidaria*.

nematocyst The thread-like contents of a *nematoblast*. See *Cnidaria*.

neo-Darwinism See *Darwinism*.

nephron or **renal tubule** The basic functional unit of the *kidney*. In mammals it consists of a cup-shaped, dilated end (*Bowman's capsule* or *renal capsule*) above which a *glomerulus* acts as an ultrafiltration unit. Glomerular filtrate enters the first part of the nephron (proximal *convoluted tubule*) which actively reabsorbs some essential materials (*glucose*, *amino acids* and approximately 80% of the water in the filtrate). The filtrate then passes into a long U-shaped tube (the *loop of Henle*), the function of which is to create a high concentration of salts in the kidney tissue. Finally, the filtrate enters the distal convoluted tubule, which is involved in the regulation of both salts and water. The permeability of the distal convoluted tubule and the collecting ducts into which its fluid is passed is variable, and determined by *antidiuretic hormone* (ADH). High ADH levels increase permeability, causing water to pass along the *concentration gradient* (created by the loops of Henle) from the tubules and ducts to the tissue fluid and blood, resulting in a concentrated urine; a watery urine is produced if ADH levels are low.

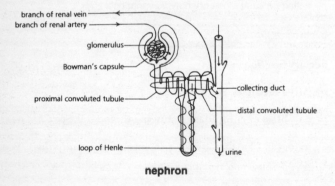

branch of renal vein
branch of renal artery
glomerulus
Bowman's capsule
collecting duct
proximal convoluted tubule
distal convoluted tubule
loop of Henle
urine

nephron

nerve A structure comprising many individual *nerve fibres*. In vertebrates, some nerves are mixed and include both sensory and motor nerve fibres, others consist of only one type of fibre.

nerve cell See *neurone*.

nerve fibre The axon or dendron, a cytoplasmic extension of a *neurone*.

nerve impulse The means by which information is carried in the nervous system, consisting of a wave of depolarization that passes along a *neurone* and is caused by changes in the permeability of the cell surface membrane that can be detected as an *action potential*. The strength (or amplitude) of the impulse is constant, only its frequency can be varied.

nerve tissue Animal tissue containing nerve cells (*neurones*) and supporting *connective tissue*; the nerve cells carry information, as *nerve impulses*, quickly from one part of the body to another.

nervous coordination Coordination of various activities with each other and with changes in the internal and external environment, carried out by a *nervous system*.

nervous system A highly organized, internal communication system found in all multicellular animals except sponges. It consists of *neurones* that interconnect sensory receptors and *effectors*. The nervous system conveys information rapidly in the form of *nerve impulses*, allowing suitable responses to internal and external stimuli to be made almost instantaneously. In vertebrates, the nervous system consists of a *central nervous system* (CNS) and peripheral nervous system. The CNS is highly complex and contains association neurones concerned mainly with processing information from different sources. The peripheral nervous system is composed of sensory and motor neurones outside the CNS, which convey information to and from the CNS respectively. Functionally, the peripheral nervous system is separated into a voluntary nervous system which controls voluntary activities (e.g. contractions of skeletal muscles), and an autonomic nervous system that regulates body activities which are usually involuntary (e.g. movements of the gut muscles in *peristalsis* and heart rate). The autonomic system has two antagonistic divisions: a sympathetic system that (like adrenaline) is generally stimulatory and prepares an animal for activity, and a parasympathetic system that tends to make an animal more relaxed.

net productivity See *productivity*.

neural arch See *vertebra*.

neurogenic Induced by nervous stimulation. The contraction of skeletal muscle is usually neurogenic.

neurone or **nerve cell** The basic unit of the *nervous system*; a cell specialized for the transmission of *nerve impulses*, consisting of a cell body and cytoplasmic extensions that may be more than a metre long. In mammals, sensory neurones carry messages from peripheral nerve receptors to the brain and dorsal part of the spinal cord. Motor neurones carry messages from the *central nervous system* to the effector organs (muscles and glands). Neurones obey the *all-or-none law* and carry nerve impulses in one direction only.

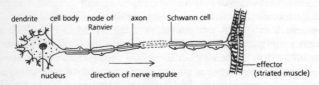

neurone A motor neurone.

neurotransmitter See *synapse*.

niacin, nicotinic acid or **vitamin B3** A water-soluble *vitamin* found in liver, fish, beans and yeast. It forms the coenzymes (see *cofactor*) *NAD* and *NADP* that play an essential role in *respiration*. Deficiency causes pellagra, the symptoms of which are dermatitis, dementia, diarrhoea, and, ultimately, death.

niche The function or role of an organism within a *community*; if the *habitat* is an organism's address, the niche is its occupation. The niche is the sum of all the physical, chemical and biological factors an organism needs to survive and reproduce. It includes the resources the organism exploits (especially food), the space it occupies (sometimes called habitat niche) and the time (e.g. day or night, summer or winter) during which it is occupying the space and using the resources.

nicotinamide adenine dinucleotide See *NAD*.

nicotinamide adenine dinucleotide phosphate See *NADP*.

nicotine A poisonous substance obtained from the leaf of the tobacco plant (*Nicotiana tabacum*); it is highly addictive.

nicotinic acid See *niacin*.

nipple See *mammary gland*.

nitrate A *salt* (NO_3^-) or *ester* of nitric acid. Metal nitrates are soluble in water. Those in the soil are essential for plant growth. They are absorbed by plants and used as their main source of nitrogen for making *proteins*. Levels of soil nitrates are often increased by the addition of artificial fertilizers. Unfortunately, up to 50% of fertilizer nitrate leaches into the ground water and can find its way into lakes and rivers where it may cause *eutrophication*. Nitrate may also find its way into drinking water. High levels of nitrate can cause a life-threatening condition in infants called 'blue-baby syndrome'. There is also evidence that high nitrate levels may increase the incidence of certain cancers. See also *nitrogen cycle*.

nitrification The oxidation of ammonia into nitrite (e.g. by *Nitrosomonas*) or nitrite into nitrate (e.g. by *Nitrobacter*) in the soil, by nitrifying bacteria which are *chemoautotrophs*.

nitrifying bacteria See *nitrification*.

nitrite A *salt* (NO_2^-) or *ester* of nitrous acid, $O=NOH$. See also *nitrogen cycle*.

nitrogen cycle The cyclical movement of nitrogen through the *biosphere*, or any subsidiary cycle which circulates nitrogen within an *ecosystem*. Organisms play a key part in the main processes involved; these include nitrogen fixation (conversion of atmospheric nitrogen into forms of nitrogen that can be used by green plants), the flow of nitrogen through *food webs* (including *ammonification*, the formation of ammonia from dead organisms by decomposers), nitrification (the bacterial transformation of nitrogen compounds into nitrites and nitrates), and denitrification (the return of nitrogen to the atmosphere by anaerobic bacteria such as *Pseudomonas*).

nitrogen fixation The conversion of atmospheric nitrogen into forms that can be absorbed by green plants. Some nitrogen is oxidized by lightning, but most is converted into ammonia by nitrogen-fixing organisms, all of which are *prokaryotes* possessing the enzyme nitrogenase. Some (e.g. the bacterium *Azotobacter*) are free-living in the soil, others have an intimate, mutually beneficial relationship with plants. An example is the bacterium *Rhizobium* that lives in *root nodules* of leguminous plants.

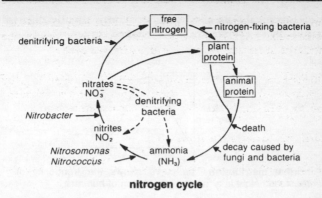

nitrogen cycle

node Part of a plant shoot to which one or more leaves are attached.

non-competitive inhibition See *inhibition*.

noncyclic photophosphorylation See *light-independent stage*.

non-essential amino acid An *amino acid* which can be synthesized by *heterotrophs* from other amino acids and which, therefore, need not form part of the diet.

non-reducing sugar A disaccharide (e.g. *sucrose*) formed by the combination of two *monosaccharides*, with the aldehyde (—CHO) group of one reacting with the aldehyde or ketone (>=CO) group of the other. Non-reducing sugars do not give a positive result in a *Benedict's test*; they need first to be hydrolysed (see *hydrolysis*) into their constituent monosaccharides by boiling in dilute hydrochloric acid.

normal distribution In statistics, a continuous distribution of data of a random variable in which the *mean*, *median* and *mode* are equal. The normal distribution is usually depicted graphically as a symmetrical, bell-shaped curve.

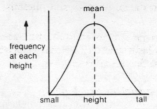

normal distribution The curve shows the frequency of recorded height in a "normal' population of humans.

nose A prominent structure at the front of the head in mammals, which acts as the organ of smell. The nose has two openings (nostrils) which lead into hair-lined passageways of the respiratory tract where air is filtered and warmed.

nostril See *nose*.

notochord A flexible skeletal rod which is present at some stage in the life of all chordates (see *Chordata*). It extends most of the length of the body, and lies beneath the dorsal nerve cord and above the alimentary canal. It supports and strengthens the body and acts as a fulcrum against which muscles can act during locomotion. In vertebrates, the notochord is replaced by the backbone.

nuclear Pertaining to the *nucleus* of a cell.

nuclear division See *meiosis* and *mitosis*.

nuclear envelope Two membranes that surround the *nucleus* of a cell. The outer membrane is continuous with the *endo-*

plasmic reticulum and may be engaged in *protein synthesis* if covered with *ribosomes*. The nuclear envelope is perforated with pores (called *nuclear pores*) that allow exchange of substances between the *cytoplasm* and nucleus.

nuclear membrane The membrane (in fact two membranes) visible under a light microscope that appears to surround the *nucleus* of a cell. See *nuclear envelope*.

nuclear pore See *nuclear envelope*.

nucleic acid A complex, long, thread-like macromolecule made of a chain of *nucleotides*. Nucleic acids contain the store of genetic information in all living things. They occur in two forms: *DNA* (deoxyribonucleic acid) and *RNA* (ribonucleic acid).

nucleolar organizing region A region of a *chromosome*, associated with a *nucleolus*, that contains a cluster of *genes* responsible for controlling the synthesis of ribosomal *RNA*.

nucleolus (*pl.* **nucleoli**) A small, dense, round body within the non-dividing *nucleus* of plant and animal cells. Some cells have more than one nucleolus. A nucleolus contains proteins, DNA and ribosomal RNA. It is involved in the synthesis of *ribosomes*.

nucleoplasm See *nucleus*.

nucleoside An organic compound consisting of a combination of a pentose sugar with a base.

nucleosome The basic form in which *DNA* is packaged in the *chromosomes* of eukaryotic cells (see *eukaryote*). Each nucleosome is bead-like and consists of a segment of DNA (146 base pairs) wound around a protein core composed of eight *histone* molecules (two copies of each of four types of histone).

nucleotide An organic compound which consists of a base, sugar and phosphate. Single nucleotides include *adenosine triphosphate*, dinucleotides include *NAD* and *NADP*. Nucleotides can combine to form the polynucleotides *DNA* and *RNA*.

nucleus (*pl.* **nuclei**) A conspicuous *organelle*, found in most *eukaryote* cells, that contains DNA, the inherited material which controls all the activities of the cell. Nuclei are typically spherical to ovoid in shape, about 10—20 μm in length, and surrounded by a *nuclear envelope*. Within the nucleus is a gel-like matrix (the *nucleoplasm*) containing one or more *nucleoli* and *chromatin*. Division of the nucleus is the basis of cell replication and reproduction (see *meiosis* and *mitosis*). Mature *red blood cells* lack a nucleus and are unable to divide.

nutrient Any substance taken in by an organism and used for growth and repair of tissues, to maintain health, or as a source of energy.

nutrient recycling The release into the environment of nutrients used by living organisms in such a way that they can be re-used by other organisms. See *carbon cycle* and *nitrogen cycle*.

nutrition The means by which an organism obtains energy and *nutrients* for any of its activities. See *autotrophic nutrition* and *heterotrophic nutrition*.

obligate parasite See *parasite*.

observation 1. The examining of the details of an event; this usually involves concentrating on particular relevant details of

the event and not simply looking at it. Observations may be qualitative or quantitative.
2. That which is observed.

oesophagus The part of the *alimentary canal* in vertebrates that leads from the *pharynx* to the *stomach*. The oesophagus is capable of moving food by *peristalsis*, but produces no digestive juices.

oestrogens A group of *hormones* in vertebrates, produced mainly by the ovary; in mammals, they are also secreted by the placenta and, in small amounts, by the *adrenal* cortex and male *testis*. In females, oestrogens help maintain *secondary sexual characteristics* (e.g. breasts) and are involved in the repair of the uterine wall after *menstruation*.

offspring or **progeny** The descendants of an organism. See also F_1 and F_2.

oil A *lipid* similar in structure and function to a *fat*. Both are triacylglycerols, but oils are liquid at room temperature (20 ℃). See also *fatty acid*.

olfaction See *smell*.

olfactory organ See *smell*.

Oligochaeta See *Annelida*.

omasum See *rumen*.

omnivore An animal that feeds on a mixed diet of animal and plant material. Compare *herbivore* and *carnivore*.

one-gene-one-polypeptide theory See *genetic code*.

oocyte A cell in the ovary which undergoes *meiosis* to form the *egg*. Cells undergoing the first meiotic division are called primary oocytes, after which they become secondary oocytes. These secondary oocytes undergo the second meiotic division to form egg cells.

open circulatory system See *circulatory system*.

operculum *1.* A lid or flap of skin covering an opening, such as the gill slit cover of bony fish and the calcareous plate on the foot of a gastropod mollusc which closes off the entrance to the shell.
2. The cone-shaped lid of a moss capsule.

optic (of structures or processes) Related to the *eye* or *vision*.

optic nerve A large nerve that leaves the back of the vertebrate eye and carries information from the retina to the visual area of the brain cortex.

oral (of parts of the body, and functions) Related to the mouth.

order See *taxon*.

organ A distinct, multicellular structure in animals, usually composed of different tissues, which is adapted for one or more particular functions (e.g. heart, kidney, liver). Some biologists also regard plant structures (e.g. leaves and roots) as organs, but others believe plants are restricted to a tissue level of organization and have no true organs.

organelle A discrete structure with a specific function found within a cell. Organelles are usually separated from the rest of the *cytoplasm* by one or more *cell membranes*; these segregate the chemicals they contain (and their reactions) from the rest of the cell.

organic evolution The *evolution* of life.

organic fertilizer See *fertilizer*.

organism A discrete, living thing that is potentially capable of carrying out all the functions of a living system (i.e. reproduction, growth, nutrition, excretion, respiration, irritability, movement).

organ of Corti A structure in the *ear*, running the length of the *cochlea*, containing sensory cells that respond to vibrations in the cochlear fluid set up by sound waves transmitted from the outer ear.

ornithine cycle or **urea cycle** A metabolic pathway in liver cells, in which ammonia, produced from the breakdown of excess *amino acids*, is combined with carbon dioxide to form *urea*.

osmoregulation The control of the *water potential* within an organism in order to maintain a constant volume of fluids. Various mechanisms of osmoregulation are used by freshwater animals to compensate for the influx of water by *osmosis* into their bodies. *Amoeba* uses a *contractile vacuole* to bale out excess water. Freshwater fish produce a watery urine and actively transport salts from the water into their bodies. Terrestrial mammals have the opposite problem of overcoming the risk of dehydration. Again, a number of strategies are adopted to maintain a water balance. Desert rats, for example, have especially long loops of Henle (see *nephron*) so that they can produce a very concentrated urine; they eat food which, when respired, releases a substantial amount of water, but during the day they inhabit burrows where their stored food can absorb the moisture exhaled with their breath. Organisms that live in water with the same water potential as their own bodies

209

have no need for osmoregulation. Some estuarine and seashore animals cannot osmoregulate, but can tolerate substantial changes in the total volume of their body fluids.

osmosis The net movement of water molecules from a region of high *water potential* (where the concentration of water molecules is high) to a region of lower water potential (where the concentration of water molecules is low), through a *partially permeable membrane*. Osmosis is a passive process and involves the movement of water molecules only.

osmotic potential See *solute potential*.

osmotic pressure The pressure required to stop the flow of a pure solvent (usually water) into a solution through a semi-permeable membrane. Osmotic pressure has a positive value. In biology, the use of the term *solute potential* is generally recommended.

Osteichthyes, teleosts or **bony fish** A class of animals (phylum *Chordata*) that have a skeleton made of bone, a terminal mouth, gills covered by a single operculum (a bony flap of tissue), and fins supported by rays of bones.

outer ear See *ear*.

oval window or **fenestra ovalis** A small membrane that allows vibrations to be transmitted from the middle to the inner *ear*.

ovary The organ that produces the female *gametes*. See also *flower*.

overfishing The removal of so many fish from a population that the population cannot maintain its numbers and goes into decline. Overfishing usually results from catching excessive numbers of small fish that are not yet mature enough to reproduce.

overpopulation The situation in which the population density of organisms exceeds the carrying capacity of their environment.

oviduct A tube that carries an animal egg cell from the ovary to the exterior or to another part of the reproductive system. In mammals, the oviduct is sometimes called the *Fallopian tube*; it conveys egg cells by the action of muscles and *cilia* from the ovary to the *uterus*.

ovulation The release of an *oocyte* from an ovary into the oviduct ready for *fertilization* by a male *gamete*. In mammals, an oocyte is released from one of the two ovaries by rupture of the *Graafian follicle* and ovary wall about once every 28 days.

ovule A structure in the carpel of flowering plants (*Angiospermatophyta*), connected to the ovary by a stalk called the *funicle*. The ovule contains an egg cell (female gamete) within an *embryosac* surrounded by integuments. The ovule develops into the *seed* after *double fertilization*.

ovum See *egg*.

oxidation Any reaction involving the addition of oxygen, removal of hydrogen, or loss of an electron. See also *redox reaction*.

oxidative decarboxylation A process that occurs during the Krebs cycle of *aerobic respiration*, when carbon atoms derived from *acetyl coenzyme A* are oxidized to carbon dioxide.

oxidative phosphorylation The process occurring in the *respiratory chain*, during *aerobic respiration*, by which *ATP* is formed from *ADP*; it is ultimately dependent on the oxidation of hydrogen by oxygen to form water. Compare *photophosphorylation*.

oxidoreductase An *enzyme* that catalyses reactions in which *oxidation* and *reduction* occur.

oxygen A colourless, tasteless gas forming about 21% of the Earth's atmosphere. It supports combustion and its availability is essential for *aerobic respiration* as it acts as the final acceptor of hydrogen and electrons in the respiratory chain. It is a waste product of the light-dependent stage of photosynthesis. The presence of oxygen can be tested with a glowing splint; if the splint relights, oxygen is present (the only other gas that can do this is chlorine, easily distinguishable because of its greenish—yellow colour).

oxygen debt A condition that occurs in aerobic animals, when more energy (*ATP*) is expended in an activity than can be generated by *aerobic respiration* and the animal resorts to *anaerobic respiration* to satisfy its energy requirements. This results in the production of *lactic acid*. When the activity is completed and the organism is at rest, extra oxygen must be consumed to oxidize the lactic acid and return the body to its pre-activity condition.

oxygen dissociation curve A curve on a *graph* showing the percentage saturation of a substance (e.g. *haemoglobin*) with oxygen against the concentration of oxygen. The oxygen dissociation curve provides information about the loading and unloading of oxygen from different respiratory pigments under different environmental conditions or in different animals.

oxyhaemoglobin The oxygenated form of *haemoglobin*; one molecule of haemoglobin can carry up to four molecules of oxygen.

oxytocin See *pituitary*.

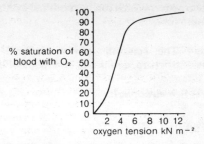

oxygen dissociation curve The characteristic S-shaped curve.

ozone A colourless gas with molecules containing three atoms of oxygen (O_3). It forms a thin layer in the upper atmosphere (between 15 and 30 km altitude) where it filters out large amounts of harmful ultraviolet light. Certain industrially made chemicals, such as chlorofluorocarbons (CFCs used as aerosol propellants, refrigerants, and foaming agents) can deplete the ozone layer (and are believed to have caused the Antarctic 'hole'). At ground level, ozone usually occurs in very small concentrations, but levels can be increased by the photochemical action of sunlight on the products of combustion of fossil fuels. At concentrations greater than 0.01 ppm, ozone is toxic and can make breathing difficult.

P A symbol commonly used in displays of solutions to genetics problems. It should be used only to designate the cross of pure-breeding (homozygous) individuals. Other types of parents are described fully as parental *phenotypes* or parental *genotypes*.

pacemaker A region of specialized muscle cells in the *sinoatrial node* of the mammalian *heart*, which initiates and maintains the heart beat. The sinoatrial node lies in the wall of the atrium near the opening of the vena cava. It has its own inherent rhythm of contraction which can be modified by hormonal and nervous stimulation.

palaeontology The scientific study of fossil organisms, including their structure, evolution, and mode of life.

palisade layer See *mesophyll*.

palisade mesophyll See *mesophyll*.

pancreas A gland that contains *endocrine* and *exocrine* tissue. The exocrine tissue secretes *pancreatic juices* into a duct that enters the *duodenum*; the endocrine tissue (*islets of Langerhans*) secretes *insulin* and *glucagon* directly into the blood stream. The pancreatic juices are a solution of alkaline salts in water which neutralize stomach acids, together with *enzymes* (e.g. lipases and amylases) or their precursors (e.g. trypsinogen) to aid digestion.

pancreatic juice See *pancreas*.

pancreomyzin or **cholecystokinin** A *hormone* secreted by the *duodenum* in some mammals, which stimulates the *pancreas* to secrete its juices and causes the *gall bladder* to eject *bile* into the small *intestine*.

Paneth cell See *crypts of Lieberkuhn*.

pantothenic acid or **vitamin B5** A water-soluble *vitamin* that forms part of *acetyl coenzyme A* and is involved in *carbo-hydrate* metabolism. Rich dietary sources include liver, yeast and peas.

parapatric speciation See *speciation*.

parapodium (*pl.* **parapodia**) See *Annelida*.

parasite An organism that lives either on (i.e. an *ectoparasite*) or in (i.e. an *endoparasite*) another living organism (the host) from which it obtains food, shelter or some other benefit. Unlike a *predator*, a parasite does not kill its prey to obtain food, but eats the tissue of its living host. The host does not benefit from the relationship (compare *mutualism*) and is sometimes harmed. Some parasites cause serious illness that may lead to death. *Facultative parasites* can exist independently as well as parasitically; *obligate parasites* can live only parasitically.

parasympathetic nervous system See *nervous system*.

parenchyma Soft plant tissue consisting of roughly spherical, relatively unspecialized cells enclosed in a thin, cellulose *cell wall*. Parenchyma forms the basic packing tissue of stems and roots. The turgidity of its cells helps support the aerial parts of the plant.

parental care All activities directed by an animal towards the protection and maintenance of its own offspring. Organisms that provide a high amount of parental care usually produce few offspring, but each one has a higher chance of surviving than an offspring which receives little parental care.

parthenogenesis A form of asexual reproduction (sometimes called virgin birth) in which female *gametes* develop into new individuals from unfertilized egg cells. The offspring are always identical with the parent and with each other. They may be haploid if the ovum has been produced by *meiosis* and diploid if it has been produced by *mitosis*. Parthenogenesis occurs in dandelions and aphids.

partially permeable membrane, differentially permeable membrane or **selectively permeable membrane** A membrane that is permeable to water but to only certain solutes. Generally, large molecules (e.g. proteins) cannot pass through a partially permeable membrane. *Cell membranes* are partially permeable. Membranes that are permeable only to solvent molecules (usually water) are called semi-permeable membranes. See also *active transport*, *diffusion*, *facilitated diffusion* and *osmosis*.

partial pressure The pressure exerted by each component in a mixture of gases. For example, if the total pressure of atmospheric gas (air) is 760 millimetres of mercury (mmHg) and air consists of 21% oxygen, the partial pressure of oxygen will be 760 x 0.21 = 160 mmHg.

pasteurization A process in which the bacterial content of certain foods and drinks (e.g. milk and wine) is reduced substantially by heating to 62—65 °C for 30 minutes. Heating destroys some bacteria, including those which cause tuberculosis.

patella or **kneecap** A small bone at the front of the knee joint which protects internal structures from damage and improves the leverage of the thigh muscles acting at the joint.

pathogen A *microorganism* (bacterium or *virus*) that causes **disease** in the plant or animal it infects.

pectoral Relating to structures in or near the chest (thorax).

Pelecypoda See *Mollusca*.

pelvic girdle, hip girdle or **pelvis** A bony or cartilaginous structure in vertebrates, to which the hind limbs are attached. The pelvic girdle articulates dorsally with the backbone.

pelvis See *pelvic girdle*.

penicillin An *antibiotic* produced by the mould *Penicillium notatum* and related species. It inhibits the synthesis of bacterial cell walls and causes bacterial cells to burst because of the inflow of water by *osmosis*.

penis The male reproductive organ, used to introduce sperm into the female reproductive tract so that internal *fertilization* can take place.

pentadactyl limb A limb, characteristic of all tetrapods, that terminates in five digits, or one that has evolved from an ancestral form possessing five digits. A pentadactyl limb typically has three parts: an upper part containing a single long bone (humerus in upper limb, femur in lower limb); a middle part (forearm or shank) containing two long bones; and a foot or hand containing a number of small bones.

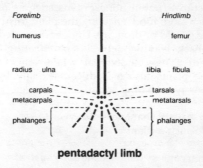

pentadactyl limb

pentose A five-carbon sugar such as ribulose (e.g. in *ribulose biphosphate*), ribose (e.g. in *RNA*), and deoxyribose (e.g. in *DNA*).

pepsin An *enzyme* formed from an inactive precursor (pepsinogen) secreted by the stomach lining. Hydrochloric acid converts pepsinogen to pepsin, which catalyses the breakdown of *proteins*.

pepsinogen See *pepsin*.

peptidase An **enzyme** that catalyses the breakdown of *proteins* by splitting off one or two amino acid units at the end of a *polypeptide chain*.

peptide bond A —CONH— bond that joins together two *amino acids*. Peptide bonds are formed by a *condensation* reaction between a carboxyl group (—COOH) of one amino acid, and an amino group (—NH$_2$) of another.

perennation The survival of an organism from one year to the next. In some *herbaceous plants*, perennation depends on special structures (perennating organs such as *bulbs*, *corms*, *tubers*), which store food, enabling the plant to survive hostile conditions (e.g. cold or drought). Under adverse conditions the plant dies back leaving only the perennating organ. This develops into a new individual when suitable conditions return. Some plants also reproduce asexually by producing more than one perennating organ.

perennial 1. A plant that lives from year to year.
2. Relating to such a plant.

peripheral nervous system See *nervous system*.

peristalsis Waves of contraction that pass along a tubular structure. Peristalsis of the intestinal wall, which pushes food along the *gut*, results from the rhythmical contraction of longitudinal and circular muscles. The longitudinal muscles elongate a section of the gut and narrow the gut *lumen*; the circular muscles shorten a section of gut and widen the gut lumen.

permeability The ability of a membrane or other barrier to allow substances to pass across it. See also *partially permeable membrane*.

pernicious anaemia See *anaemia*.

pest An organism that is harmful to humans economically or medically, or is aesthetically undesirable (e.g. garden weeds); what is considered a pest in one context may be beneficial in another. Many organisms commonly regarded as pests are harmful only when their populations increase rapidly. Pests may be controlled by one or more methods. Biological control (*biocontrol*) usually uses a natural predator of the pest (for example, ladybirds may be introduced to control aphids) in an attempt to establish a natural balance between pest and *predator*; it may also involve the release of sterilized males of the pest species, whose matings result in infertile eggs (a method used to control screw-worm fly in the USA). *Chemical control* uses poisonous manufactured substances (pesticides) to kill pests. Ideally, a pesticide should be inexpensive, effective in small doses, should harm only the target species, and should be broken down in the environment quickly, leaving no harmful residues. Some pesticides affect beneficial as well as pest organisms and persist in the external environment or in food chains (see *bioaccumulation*). Pests can also be controlled by such cultural methods as *crop rotation*, tillage, weeding, and intercropping (growing two or more crops together). The Food and Agriculture Organization of the United Nations (FAO) recommends *integrated control*, or more precisely, *Integrated Pest Management* (IPM). This utilizes all suitable pest control methods (biological, chemical and cultural) to maintain pest populations at levels below which they are harmful in ways that cause the minimum of harm to the environment.

pesticide Any substance that kills a *pest*, especially a toxic chemical compound.

petal See *flower*.

petiole The stalk by which a leaf is attached to the main stem of a plant.

pH A numerical scale ranging from 0 to 14 that indicates the acidity or alkalinity of a solution. $pH = \log_{10}(1/H^+)$ or $pH = -\log_{10}(H^+)$, where H^+ is the hydrogen ion concentration. At 25 °C, all solutions at a pH less than 7 are acidic; those at pH 7 are neutral; and those above 7 are alkaline. The pH of a solution can be measured accurately using pH meters, or approximately by using indicator papers that change colour according to the pH. Blue litmus paper, for example, turns red in an acid solution, and red litmus turns blue in an alkaline solution. Universal indicator solution (a mixture of plant and lichen pigments) gives a definite colour change at specific pH-values.

phagocytosis See *endocytosis*.

pharynx The passage leading from the mouth to the *oesophagus* in vertebrates.

phenotype The *characteristics* of an individual organism, usually resulting from interactions between the *genotype* and the environment in which development takes place. Some characteristics (e.g. eye colour) are visible in the outward appearance of the individual, but others (e.g. blood group) are internally expressed and cannot be seen. Compare *genotype*.

phloem Tissue that is specialized for transporting nutrients and plant *hormones* throughout a vascular plant. In stems of flowering plants, the phloem forms the outer part of the *vascular bundles* and consists of sieve tube elements, companion cells, and cells such as phloem fibres, which provide support. Sieve tube elements have no nucleus and are thin-walled. They are joined end-to-end by their sieve plates. These are perforated to allow continuity between one element and the next, forming a continuous channel, the sieve tube, through which sap flows. The sieve tube cannot function without the adjacent companion cells. These are small cells with a **nucleus** and other cell *organelles* lacking in the sieve tube. The companion cell controls the activity of the sieve tube element. Companion cells are linked by *plasmodesmata* (open channels in the cell walls through which strands of *cytoplasm* can flow).

phloroglucinol A chemical used to stain *lignin* red when it is acidified with hydrochloric acid.

phosphoglyceraldehyde See *glyceraldehyde 3-phosphate*.

phosphoglyceric acid See *glycerate 3-phosphate*.

phospholipid A *lipid* combined with a phosphate group. The commonest forms consist of *glycerol* attached to two *fatty acid* chains and a phosphate (PO_4^{3-}) group. Phospholipids are amphipathic, that is, they have both hydrophilic ('water-loving') and hydrophobic ('water-hating') components. The polar head, consisting of glycerol and phosphate, is hydrophilic; the non-polar fatty-acid tails are hydrophobic. Phospholipids are major components of *cell membranes*.

phosphorus A non-metallic element that is essential for life. It usually occurs as phosphates which are incorporated into a number of chemicals including *phospholipids*, *nucleotides* (e.g. *ATP*, *NAD* and *NADP*), and *nucleic acids*. In mammals, phosphates are important constituents of tissues (especially bones and teeth). Phosphate deficiency causes plant leaves to become dark green or blue—green; growth is usually reduced. Phosphorus is a relatively uncommon element and is often a *limiting factor* in the *productivity* of *ecosystems*. Plants absorb phosphates from the soil, fresh water, or the oceans, and assimilate the phosphorus into compounds that are passed along the *food chain*.

photoautotroph An organism that uses light energy from the Sun for the synthesis of complex organic molecules from inorganic materials (usually carbon dioxide). All green plants and green and purple bacteria are photoautotrophs. See *autotrophic nutrition*.

photochemical smog See *smog* (2).

photoheterotroph See *heterotrophic nutrition*.

photolysis The splitting of water into hydrogen ions (protons) and hydroxyl ions by the action of light. See also *light-dependent stage*.

photoperiodism The response of an organism to the relative duration of day and night (the *photoperiod*). Many activities exhibit photoperiodism (e.g. sexual cycles, dormancy and migration in animals; seed germination and leaf fall in plants), but the term is commonly associated with flowering in plants. There appear to be three main types of plant: those such as lettuce and clover, which require a night length of no more than a certain maximum duration; those such as strawberries, which require a night length of more than a minimum duration; and day-neutral plants which are unaffected by photoperiod. *Phytochromes* enable plants to monitor the relative length of daylight and darkness in each 24-hour cycle, and *florigen* (see *plant growth substance*) is believed to induce flowering at the appropriate photoperiod.

photophosphorylation The conversion of *adenosine diphosphate* and inorganic phosphate to *ATP*, using light energy from the Sun in *photosynthesis*. See also *light-dependent stage*.

photosynthesis The synthesis of complex organic chemicals using light energy from the Sun. Photosynthesis usually uses carbon dioxide as the source of carbon, but some purple non-sulphur bacteria use organic sources of carbon. In green plants, photosynthesis takes place in *chloroplasts* and involves two distinct stages: the *light-dependent stage* and the *light-independent stage*.

photosynthetic pigments Pigments that can absorb light energy and convert it to chemical energy. They are of two types: *primary pigments* and *accessory pigments*. When a primary pigment absorbs light, the pigment emits electrons that drive the reactions of *photosynthesis*. When an accessory pigment absorbs light, it passes electrons to the primary pigment. *Chlorophyll a* is the only primary pigment in green plants. Accessory pigments include other chlorophylls, carotenoids (orange pigments), and xanthophylls (yellow pigments).

phototaxis See *taxis*.

phototropism See *tropism*.

phylum The second highest rank or group (below *kingdom*) used in *classification*; in many systems the term is confined to animals, the plant equivalent being a division.

physiological homeostasis See *homeostasis*.

physiology The study of the internal functions and processes of organisms.

phytochrome A pale blue, light-sensitive pigment in plant leaves which exists in two interconvertible forms: Pr (or P_{660}) and Pfr (or P_{725}). Pr is converted into Pfr when it absorbs red light with a wavelength of 660 nm. Pfr is converted into Pr

molar The teeth at the back of the *jaw* in adult mammals, used for crushing. They are not preceded by *milk teeth*. Molars usually have several roots. The biting surface varies; in *carnivores* it is adapted for crushing and cutting (see *carnassial teeth*), in *herbivores* upper and lower molars have a complex pattern of projections and ridges which fit together, permitting the efficient grinding of plant material.

Mollusca or **molluscs** A phylum of invertebrate animals showing *bilateral symmetry*, most of which have an unsegmented body consisting of a head, foot and visceral hump (a soft mass of tissue containing the digestive organs, covered by soft skin). Molluscs include garden snails and slugs (class Gastropoda) that have a single calcareous shell; cockles and mussels (class Pelecypoda) which have two shells; and squids and octopuses (class Cephalopoda) which have a shell which is usually reduced and internal, and a well-developed head containing a complex brain and sense organs.

molluscs See *Mollusca*.

Monera A kingdom, in some systems of classification, that includes all *prokaryotes*: i.e. bacteria and cyanobacteria (formerly known as blue-green algae).

monoclonal antibody An *antibody* produced by *clones* of cells formed by the fusion of a particular antibody-producing B-*lymphocyte* and a cancer cell. The cell resulting from the fusion (called a *hybridoma*) reproduces by *mitosis* to form a population of genetically identical cells that make identical antibodies. Monoclonal antibodies are used to diagnose allergies and diseases (e.g. rabies and hepatitis) and it may be possible to use them to deliver toxic drugs to specific cells (e.g. cancer cells).

monocots See *Monocotyledonae*.

when it absorbs far-red light (wavelength about 725 nm) or when light is absent. During the day, Pfr builds up relative to Pr, because daylight has a higher proportion of red light. During the night Pfr is gradually converted to Pr. Thus the relative amounts of the two forms of phytochrome enable a plant to monitor day length. The interconversions can act as a switch mechanism controlling physiological processes (e.g. flowering and germination) that are linked to the relative periods of light and darkness in a 24-hour cycle. See also *photoperiodism*.

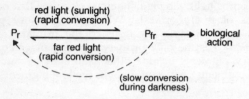

phytochrome The photoconversion of the two forms of phytochrome.

phytohormone See *plant growth substance*.

phytoplankton See *plankton*.

pie chart A diagram used to compare various quantities (usually percentages or proportions of an identifiable whole). It consists of a circle ('pie') marked into slices corresponding in size with the various quantities to be compared. See appendix 1.

pigment A coloured substance. Pigments have a number of rôles in living organisms. They include cytochromes that act as *electron carriers* for *respiration*; *chlorophyll*, involved in photosynthesis; and *haemoglobin* which transports oxygen in mammals. See also *photosynthetic pigments*.

pili or **fimbriae** Sticky protein rods extending from the surface of some bacteria. Pili are involved in cell-to-cell or cell-to-substrate attachments.

pinna The outer, funnel-shaped flap that directs sound waves into the *ear*, in mammals.

pinocytosis See *endocytosis*.

pitfall trap A device consisting of a container such as a glass jar or disposable cup, sometimes containing bait or preservative, sunk into the earth to trap invertebrates as they move over the ground surface. Usually a cover, raised above the surface, is placed over the trap to prevent the entry of rain or animals such as frogs and toads which may feed on the trap contents.

pituitary An *endocrine gland* at the base of the *brain* in vertebrates, sometimes referred to as the 'conductor of the endocrine orchestra', because of its role in coordinating the activity of other endocrine glands; many of its activities are controlled by the hypothalamus of the brain. The anterior part secretes many *hormones* which regulate other endocrine glands (e.g. *follicle stimulating hormone*, and thyroid stimulating hormone which controls the activity of the *thyroid gland*). Hormones secreted by the posterior part include *antidiuretic hormones* and oxytocin (which, in mammals, stimulates contraction of *smooth muscle* during birth and of parts of the mammary glands during suckling).

placenta 1. (in mammals) A temporary, spongy, double-layered structure, formed from the tissues of *embryo* and mother, by which the embryo is attached to the *uterus* wall during pregnancy. In the placenta, the close association between the blood vessels of the mother and the embryo allows nutrients and oxygen to pass to the embryo and carbon dioxide and metabolic wastes to be removed from it, but there is no direct connection between maternal and embryonic blood. The placenta also produces various *hormones* which help maintain pregnancy.

2. (in flowering plants) The part of the ovary wall to which the *ovule* is attached.

plagioclimax See *climax*.

plankton Aquatic organisms that drift with water movements. Most are small and have limited mobility. Phytoplankton consists mainly of unicellular *autotrophs* that form the basis of many aquatic food chains. Zooplankton includes *heterotrophic* unicells and multicellular organisms (e.g. jellyfish).

Plantae or **plants** The kingdom comprising all multicellular, eukaryotic (see *eukaryote*), photosynthetic organisms with cells containing *chloroplasts* and surrounded by a cellulose cell wall. A typical plant is *autotrophic* and non-motile.

plant growth substance, phytohormone or **plant hormone** One of a group of chemicals produced inside a plant that plays an important role in the coordination of plant growth and development. Most plant responses are slow and involve the differential growth of tissues; plant growth substances control most of these responses, being substances which, in very small concentrations, can increase, decrease, or otherwise change the growth of a part or whole of a plant. They should not be called hormones, because they do not always move from their sites of production. *Auxins* (the most common is indoleacetic acid, IAA, synthesized in shoot tips) promote shoot growth by stimulating cell elongation and differentiation. Auxins may concentrate down one side of the plant to induce localized growth which, for example, results in a stem growing towards light (see *phototropism*). Auxins inhibit elongation of root cells except at relatively low concentrations and inhibit growth of lateral shoots, thus promoting apical dominance. Synthetic auxins (e.g. 2,4-D and 2,4,5-T) are used as weedkillers. *Cytokinins* (also known as *kinins*) work with IAA to affect the rate of cell division, promoting bud formation, and are essential for the growth of healthy leaves. *Gibberellins* (e.g. gibberellic acid), produced in the apical leaves, buds, seeds and root tips, cause cells in the stem to elongate and may cause leaves to increase their area; they can be used to induce bolting (unusual lengthening of plant stems) in immature plants and can remove dwarfness in mature plants. They are also thought to play a role in flowering, seed germination, and the development of fruit. *Abscisic acid* (also known as abscisin or *dormin*) is synthesized mainly by *chloroplasts*, but is found throughout plants, especially in the leaves, fruits and seeds. It has powerful growth-inhibiting properties and is involved in leaf fall; it is believed to cause *stomata* to close during periods of drought, and may inhibit other plant growth substances by preventing protein synthesis. *Ethene* (*ethylene*) is produced by most plant tissues during normal plant metabolism and may cause leaves to fall and fruits to ripen. *Florigen* is believed to promote the development of flowers, but it has not yet been isolated. Many plant responses result from the interaction of a number of plant growth substances. See also *phytochrome*.

plants See *Plantae*.

plaque A thin layer of hard material that may cover all or part of a tooth. It consists of food (mainly sugars) and bacteria. The bacteria digest the food and produce acids which contribute to tooth decay unless neutralized.

plasma See *blood*.

plasma membrane See *cell membrane*.

plasmid A small, circular piece of *DNA* found in bacteria and some yeasts. Because plasmids can replicate independently of the *chromosome* and may pass from one cell to another they are used in *genetic engineering* to transfer DNA from one organism to another.

plasmodesmata See *phloem*.

plasmolysis The shrinkage of a plant cell so that the *plasma membrane* no longer presses against the *cell wall*. Plant cells plasmolyse when they are in a solution that has a lower *water potential* than exists inside the cell. The stage at which the plasma membrane begins to lose contact with the cell wall is called *incipient plasmolysis*. A cell at this exact stage is impossible to distinguish from a turgid cell. For practical purposes, therefore, the point of incipient plasmolysis is often assumed to occur when 50% of the cells are plasmolysed.

platelet See *blood*.

Platyhelminthes or **flatworms** A phylum comprising flat, unsegmented animals that are *triploblastic* and lack a *coelom*. Most (except tapeworms and other cestodes) have a mouth but no anus. They are usually *hermaphrodite* with a complex reproductive system.

plumule The part of an *embryo* in a seed that develops into the shoot.

poikilotherm or **exotherm** An animal (e.g. a jellyfish) that cannot regulate its body temperature which, therefore, varies according to the temperature of its surroundings. Poikilotherms are sometimes called 'cold-blooded', but their body temperatures may be high as well as low. Some organisms (e.g. certain fish and reptiles) which were previously classified as poikilotherms are now known to regulate their body temperatures within a fairly narrow band. Compare *ectotherm*. See also *homeotherm*.

point quadrat A device for sampling stationary organisms (usually plants) at points, consisting of a pin-frame (legs of adjustable height supporting a cross bar with holes down which pins, usually 10, are inserted). The quadrat is placed on the ground at a randomly selected site with all the pins withdrawn; each pin is lowered and a note made of all the organisms penetrated by it, so percentage cover can be estimated. If vegetation of variable height is being sampled, records of top cover (first plants penetrated by the pin) and bottom cover may be taken. See *quadrat*.

pollen grain A small spore (microspore), produced in large numbers in the anthers of flowering plants by *meiosis* of *pollen mother cells*. Each pollen grain consists of a tough outer coat (*exine*) and more delicate inner coat (*intine*) enclosing two generative nuclei involved in *double fertilization*, and a vegetative nucleus responsible for directing the growth of the *pollen tube*. The contents of the pollen grain represent the male *gametophyte*. Pollen grains of insect-pollinated plants are often spiny or pitted. Those of wind-pollinated plants are often small, smooth and light.

pollen mother cell One of the *diploid* cells which line the pollen sacs in the anthers of flowering plants. They undergo *meiosis* to form *pollen grains*.

pollen tube An outgrowth of a *pollen grain*, in flowering plants. It provides a protective, watery channel through which male *gametes* can enter the ovule for *double fertilization*. A pollen tube will penetrate an ovule only if it is genetically compatible with the female tissue.

pollination The transfer of a *pollen grain,* in flowering plants, from an anther to a stigma, either within the same flower or between flowers of the same plant (self-pollination), or between flowers of different plants (cross-pollination). Cross-pollination requires the action of a pollinating agent such as wind, insects or water.

pollution The introduction into the environment of substances or energy that are not naturally present, or not in such large concentrations, and that may cause harm to humans, other living organisms valued by humans, or the environment itself. Pollution may be physical (e.g. noise, and heat and other forms of radiation), chemical (e.g. pesticides), or biological (e.g. sewage). See also *acid rain, eutrophication* and *greenhouse effect.*

Polychaeta See *Annelida.*

polygene See *polygenic inheritance.*

polygenic inheritance The inheritance of a phenotypic characteristic (e.g. height and intelligence in humans) that is determined by more than two *genes,* each of which has only a small effect. The group of genes that collectively affects a characteristic is called a *polygene.* Characteristics inherited in this way usually show continuous variation and are strongly affected by the environment.

polymer A large molecule formed by the linking together of a number of smaller molecules (monomers). *Protein* is a polymer made of *amino acid* subunits; *starch* is a polymer made of *glucose* subunits.

polynucleotide A chain of *nucleotides. Nucleic acids* are polynucleotides. See *DNA* and *RNA.*

polypeptide chain A chain of 10 or more *amino acids* joined together by *peptide bonds*. The sequence of amino acids in a single polypeptide chain is determined by the sequence of bases in part of a *DNA* molecule that represents one gene. *Protein* consists of one or more polypeptide chains usually of more than 100 amino acids. Certain *antibiotics* and *hormones* consist of polypeptide chains with fewer amino acids.

polyploidy The condition in which the cells of an organism have more than two sets of *chromosomes*; for example, commercial wheat is hexaploid, i.e. it has six sets of chromosomes.

polysaccharide A long chain of simple sugar molecules joined together. Polysaccharides are relatively insoluble, not sweet-tasting, and have a relatively small effect on the osmotic potential of a solution. The commonest polysaccharide is *cellulose*; others include *starch* and *glycogen*.

pooter A device for collecting small terrestrial animals. It acts like an aspirator, the animals being drawn through a tube and into a chamber by sucking at a mouthpiece, the inner end of which is covered with muslin. A *blow pooter* is used for collecting animals from unpleasant surfaces (e.g. cow dung). With this device, blowing down a mouthpiece creates air currents which suck the animal into the chamber.

population 1. A group of organisms, all of the same species, which occupies a particular area.
2. The total number of individuals of a given species or other taxonomic group in a defined area.

population control The limitation by humans of the number of organisms within a population, commonly a human or pest population.

population dynamics The study of population fluctuations, i.e. of how the number of individuals belonging to one or more species changes with time and how environmental factors effect those changes. See also *density-dependence* and *density-independence*.

population growth 1. An increase in the number of individuals in a population that occurs when the birth rate exceeds the death rate.
2. Changes over time in the number of individuals in a population; this can be shown graphically as a growth curve, of which there are two main types: a *J-shaped curve* for *r-species* and an S-shaped curve for *K-species*.

positive feedback See *feedback*.

post-anal tail An extension of the backbone that extends beyond the *anus*. A characteristic feature of most *chordates*, it may be used for locomotion, for balance, or for gripping objects.

posterior (of parts of an animal body) At or near the back or rear end; usually the region directed backwards when an animal is moving. In humans, it is synonymous with *dorsal*.

potential energy Stored energy; i.e. energy capable of doing work, but not presently doing so (e.g. chemical energy, and the mechanical energy possessed by an object because of its position).

potometer An instrument used to measure the rate at which a cut plant shoot absorbs water. The base of the shoot is immersed in a water reservoir and the absorption rate is indicated by the time taken for either a small air-bubble, or the water meniscus, to move a stipulated distance along a capillary tube. Assuming a constant ratio between the rate at which water is absorbed and the rate at which it is lost by transpiration from the exposed tip (or leaf), the rate of transpiration can be calculated from the potometer reading.

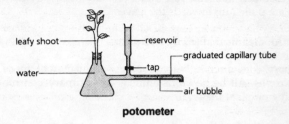

potometer

predator A *heterotroph* that kills another organism (its prey) in order to obtain nourishment from it.

pregnancy The condition, in mammals, that follows successful *fertilization* of an egg cell and its *implantation* on the *uterus* wall. The duration of pregnancy (*gestation period*) extends from fertilization to birth and during it the life of the *embryo* is maintained by the mother, mainly by exchange of materials via the *placenta*, while *hormones* (e.g. *progesterone*) prevent further *ovulation*.

premolar One of the teeth in mammals between the *canines* and the *molars*. In herbivores, premolars are usually ridged and furrowed and used for crushing; in carnivores, they may be adapted for shearing (see *carnassials*).

pressure potential, ψP, turgor pressure or **wall pressure** The hydrostatic pressure to which water in the liquid state is subjected. In a turgid plant, water inside the cell produces a hydrostatic pressure which pushes against the cellulose *cell wall*; the

233

wall responds by producing an equal pressure in the opposite direction. This pressure is sometimes referred to as wall pressure and is usually positive. During *transpiration*, however, the pressure potential within *xylem* vessels is usually negative because water molecules are under tension (being stretched) as they are pulled up towards the leaves.

prey An organism killed by a *predator*.

primary consumer A heterotrophic organism that feeds on autotrophs. All herbivorous animals are primary consumers, feeding on plant material.

primary pigment See *photosynthetic pigments*.

primary producer or **producer** An organism that manufactures food from simple inorganic substances by processes such as *photosynthesis*. Primary producers (e.g. all green plants) form the first **trophic level** in food pyramids.

primary productivity See *productivity*.

primary sexual characteristic A feature of an organism that is directly concerned with producing the *gametes*.

primary structure The linear sequence of *amino acids* within a *polypeptide chain* of a *protein*.

primary succession See *succession*.

probability The statistical chance or likelihood that a given event will occur. Probability is expressed as a value between 0 (complete certainty that an event will not occur) and 1 (complete certainty that an event will occur) or as a percentage between 0% and 100%.

problem A phenomenon, situation or question that is difficult to deal with or requires explanation. In science, problems may arise from a practical need (e.g. to prevent bacterial contamination of milk) or from scientific observations.

producer See *primary producer*.

product A substance produced as a result of a chemical reaction. Compare *substrate*.

productivity The rate at which *biomass* is produced or energy assimilated in an *ecosystem* or part of an ecosystem. There are two components: primary productivity (the production of new organic matter or assimilation of energy by *autotrophs*) and secondary productivity (the production of new organic matter or assimilation of energy by *heterotrophs*). Each component can be divided into gross productivity (the total amount of organic matter produced or energy assimilated) and net productivity (the organic matter or energy used for growth, excluding that used for *respiration*).

progeny See *offspring*.

progesterone A *hormone* secreted mainly by the *corpus luteum* in the ovaries of mammals, but also by the *placenta* and, in small amounts, by the male *testis*. Its main function appears to be preparation of the *uterus* for *implantation*. It also prevents *ovulation* during *pregnancy* and assists in the development of the placenta and mammary glands.

Prokaryotae A kingdom comprising all *prokaryotes* (in some classifications).

prokaryote An organism characterized by the lack of a *nuclear envelope*, lack of double-membraned *organelles* (e.g. *chloro-*

plasts and *mitochondria*), and lack of organelles with *micro-tubules* that have a 9 + 2 arrangement (see *cilium*). Typically, prokaryotes are unicellular or filamentous organisms that have circular DNA lying naked in the *cytoplasm*. See also *bacterium*.

propagation 1. The transmission of a *nerve impulse* as a wave of depolarization along a *nerve fibre* without loss of strength. **2.** Asexual or sexual plant reproduction.

prophase The first main stage in nuclear division, during which *chromosomes* condense and then (in *mitosis* and *meiosis* I) sister *chromatids* appear while the nuclear membrane and nucleoli disappear. In the prophase of diploid cells, *homologous chromosomes* behave independently during mitosis, but in meiosis I they associate into *bivalents*, with chromatids *crossing over*. At the end of prophase, the homologous pairs separate and move towards the equator of the *spindle apparatus*.

proprioceptor An internal sensory receptor that is sensitive to movement, pressure, or stretching of structures within the body. In mammals, proprioceptors are usually located in muscles, tendons and ligaments, and provide information essential for smooth, coordinated movement and maintenance of body posture.

prostate gland A gland in male mammals that surrounds and opens into the urethra near the bladder. Secretions from the prostate gland include alkaline substances that activate sperm and prevent them from sticking together. These secretions are added to the **semen** before ejaculation.

prosthetic group A non-protein group attached to a *protein* molecule so tightly that it cannot be removed by *dialysis*. A prosthetic group may be inorganic (e.g. a metallic ion) or organic (e.g. the iron-containing haem groups in *haemoglobin*). In *enzymes,* a prosthetic group is often attached to the *active site* and is essential for the enzyme's activity.

protease A digestive *enzyme* (e.g. pepsin or trypsin) that catalyses the breakdown of *proteins*.

protein A complex organic compound always containing carbon, hydrogen, oxygen and nitrogen, and sometimes sulphur. Proteins are essential components of all living organisms. Each protein molecule is made of one or more *polypeptide chains* consisting of 100 or more *amino acids* linked together by *peptide bonds*. In conjugated proteins, non-protein groups are attached to a polypeptide chain. Proteins are manufactured on *ribosomes* using instructions carried by *messenger RNA*. Each protein has a unique, genetically determined structure. There are four levels of protein structure. The primary structure is the linear sequence of amino acids. The secondary structure is the spatial arrangement of groups of amino acids in a polypetide chain. There are two main types of secondary structure: the *alpha helix* and the *beta-pleated sheet*. The tertiary structure is the overall shape of a single polypeptide chain and depends on how the chain is folded to form a three-dimensional structure. When heated or subjected to strong acids or alkalis, proteins lose their specific tertiary structure and become denatured. The quaternary structure is the three-dimensional arrangement of all the polypeptide chains in a protein. For example, *haemoglobin* has four polypeptide chains, each of which changes shape slightly when an oxygen molecule attaches to its iron-containing group. Proteins can be broadly classified as either fibrous or globular. Fibrous proteins are usually insoluble and have a structural role. Globular proteins are usually soluble and play a role in *metabolism*. All *enzymes* are globular proteins. Plants synthesize proteins using *carbohydrates* from photosynthesis and *nitrates* which they absorb from soil or water. Animals obtain their proteins from plants. Protein can be used as a *substrate* for *respiration*, especially during periods of starvation or extreme exertion; each gram of protein yields about 17 J (4 calories) of energy. The presence of proteins in a solution can be demonstrated using the *Biuret test*.

protein synthesis An *anabolic reaction* in which a *protein* is formed from *amino acids*. In *eukaryotes*, protein synthesis is under the control of a section of DNA (the *cistron*), the base sequence of which determines the amino acid sequence in a *polypeptide* chain of the protein. DNA is transcribed (see *transcription*) to mRNA which moves from the nucleus into the *cytoplasm*, where it associates with *ribosomes*; these translate (see *translation*) the *genetic code* embodied in the mRNA base sequence into a polypeptide chain. Different tRNA molecules are involved in carrying specific amino acids to the ribosomes and attaching them to the polypeptide chain at the appropriate point.

prothrombin See *blood clotting*.

Protoctista A *kingdom* comprising organisms in which the cells have an organized nucleus surrounded by a *nuclear envelope* (see *eukaryotes*), but which are not animals, plants or fungi. The Protoctista includes *unicellular organisms* and nucleated *algae*.

protoplast The whole of a plant cell, apart from the *cell wall*; the living part of a plant cell comprising a *nucleus* and *cytoplasm*. Protoplasts are sometimes isolated by removing the cell wall so that metabolic processes can be more easily studied.

proximal (of a structure) nearest to the midline of the body, or closest to its point of attachment to the body.

pseudopodium (*pl. pseudopodia*) Literally, a 'false leg'; a temporary extension of the *cytoplasm* that enables *Rhizopoda* (e.g. *Amoeba*) and *white blood cells* to move and feed. See also *amoeboid movement*.

pterygote See *imago*.

puberty The beginning of sexual maturity in humans. In females, it begins after the first loss of blood associated with *menstruation*. In males, it begins with the first development of *sperm*.

pulmonary Pertaining to the lungs.

pulp See *tooth*.

pulse 1. The rhythmic expansion and recoil of the arteries resulting from the wave of pressure produced by contraction of the left ventricle of the *heart*. The pulse can be felt in any artery which passes over a bone and is close to the body surface.
2. The edible seed of a leguminous plant (e.g. pea, bean, lentil).

Punnett square A table for demonstrating all the possible *genotypes* that can result from the random fusion of *gametes*, first used by the Cambridge geneticist R.C. Punnett. The alleles within the gametes of one parent are written along the top of a series of boxes, and the alleles within gametes of the other parent are written along the side. The products of random fusions are written in appropriate boxes from which the proportion of each genotype and *phenotype* can be estimated. Sometimes a diamond-shaped checkerboard is used instead of the traditional Punnett square.

	♀ gametes			
	AB	Ab	aB	ab
♂ gametes aB	AaBB	AaBb	aaBB	aaBb
ab	AaBb	Aabb	aaBb	aabb

Punnett square

pupa The stage between larva and adult, in the life cycle of insects which undergo a complete *metamorphosis*. Pupae are often inactive and enclosed in a hard, protective case of shelly material (*chrysalis*) or silky material (*cocoon*). Feeding ceases and large changes occur, associated with becoming an adult.

pupil See *eye*.

pure breeding, breeding true or **pure line** (of organisms) Homozygous for a pair of alleles of a particular *gene* or genes. Pure-breeding organisms that are self-fertilized or mated with each other produce offspring identical to the parent for the genes being considered (unless a *mutation* occurs); if the offspring mate with each other they produce a new generation of identical offspring, and so on for successive generations.

pure line See *pure breeding*.

purine An organic nitrogen base. Purines have a double-cyclic (double ring) structure. They include adenine and guanine, which occur in *DNA* and *RNA*.

pyloric sphincter A ring of *smooth muscle* in vertebrates, which, when contracted, closes the junction between the *stomach* and *duodenum*, thus controlling the movement of *chyme* from stomach to duodenum.

pyramid A conical structure or protruberance (e.g. in the medulla of the *kidney*).

pyramid of biomass A diagrammatic representation of the *biomass* (usually expressed as total dry mass of living matter) in each *trophic level* of an *ecosystem*. Typically, each trophic level is represented by a rectangle, the length of which is proportional to its biomass. Starting with the rectangle for the first trophic level (primary producers) as the base, each subsequent rectangle is drawn on the centre of the preceding rectangle. A similar diagram can be drawn to show the number of individuals at each trophic level (*pyramid of numbers*) or the flow of energy from one trophic level to the next (*pyramid of energy*). The diagrams may assume markedly different forms, but the shape of the diagram showing the flow of energy assumes that of a pyramid, because at each trophic level some of the usable energy is lost as heat energy.

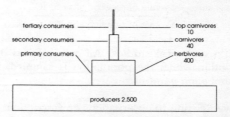

tertiary consumers — top carnivores 10
secondary consumers — carnivores 40
primary consumers — herbivores 400
producers 2,500

pyramid of biomass Hypothetical ecological pyramid of energy for the standing crop on tropical grasslands. Figures in Kcal m^{-2} yr^{-1}.

pyramid of energy See *pyramid of biomass*.

pyramid of numbers See *pyramid of biomass*.

pyrimidine An organic nitrogen base. *DNA* contains the pyrimidines thymine and cytosine; *RNA* contains cytosine and uracil. Pyrimidines have a single ring structure.

pyruvate or **pyruvic acid** A three-carbon organic acid ($CH_3.CO.COOH$) produced during *glycolysis*. In *eukaryotes*, if oxygen is present, pyruvate is converted into *acetyl coenzyme A* and fed into the Krebs cycle for *aerobic respiration*. In the absence of oxygen, it is converted to *lactic acid* in animals and ethanol and carbon dioxide in plants.

pyruvic acid See *pyruvate*.

Q_{10} or **temperature coefficient** A measure of the effect of a 10 °C rise in temperature on the rate of a chemical reaction. The Q_{10} is expressed as the ratio of the rate of a chemical reaction at a given temperature to the rate of the same reaction at a temperature 10 °C lower, i.e.

$$Q_{10} = \frac{\text{rate of reaction at x+10 °C}}{\text{rate of reaction at x °C}}$$

quadrat (in ecological studies) An area marked out on the ground within which the population is sampled. See *frame quadrat* and *point quadrat*.

qualitative Relating to observations or descriptions that have not been measured and are therefore subjective. 'Red', 'green', and 'blue' are qualitative descriptions of the colour of light.

qualitative variation See *variation*.

quantasome A particle containing *chlorophyll* that occurs on the membranes within a *chloroplast*. Quantasomes are regularly arranged and there appear to be two sizes, each with its own characteristic set of about 300 chlorophyll molecules.

quantitative Relating to observations or descriptions that have been measured (e.g. 760 nm is a quantitative description of the colour of light in terms of its wavelength).

quantitative variation See *variation*.

quaternary structure See *protein*.

radial symmetry The circular arrangement of body parts in an organism, such that the body can be divided into two halves by cutting along more than one plane running through the centre of the body. Radially symmetrical animals (e.g. sponges, sea anemones and starfish) tend to be non-moving or very slow moving, and the stimuli to which they respond (especially food) are equally likely to come from any direction.

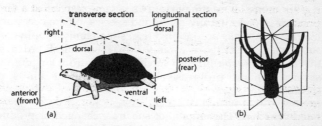

radial symmetry Comparison of bilateral symmetry (left) and radial symmetry (right.)

radiation 1. The emission or transfer of electromagnetic energy (e.g. light, infrared rays, X-rays and gamma-rays) as rays, waves or particles.
2. See *adaptive radiation*.

radicle Tissue within the *embryo* of a seed that develops into the root.

radioactive tracer A molecule or atom that has been labelled with a radioactive substance (e.g. one containing the radio-active isotope ^{14}C) that emits ionizing particles which can be detected by a Geiger—Müller counter, on an *autoradiograph*, or by some other means. The substance can thus be followed during a chemical reaction, metabolism, or passage through an organism.

radius A bone in the forelimb of a tetrapod.

random sampling Selecting an item (such as an organism or a sampling area) from a group in such a manner as to give each item present an equal chance of being selected.

rank One of the major groups into which living organisms are classified. There are seven ranks commonly used in biological classifications: kingdom, phylum (in plants, often called a division), class, order, family, genus and species. Where necessary, intermediate ranks are inserted (e.g. superfamily, subphylum, subspecies). Ranks are arranged in hierarchical order so that taxa (see *taxon*) in each successive rank from kingdom to species contain progressively fewer different kinds of organism.

rate of reaction The amount of substrate changed or product formed during a given time, in a chemical reaction. An *enzyme* reaction does not have a constant rate: if the amount of sub-strate changed or product formed is plotted against time, a curve can be drawn and the rate of the enzyme reaction determined from the slope of the tangent drawn to the curve in the initial stage of the reaction; the steeper the slope, the greater the rate.

reactant A substance that takes part in a chemical reaction.

receptacle See *flower*.

receptor cell or **sensory cell** An animal cell specialized to detect a change in the internal environment (*interoceptor*, e.g. a *proprioceptor*) or the external environment (*exteroceptor*, e.g. rods and cones in the retina) and initiate the transmission of a *nerve impulse* that may affect the behaviour of the animal. All sense organs contain receptor cells.

recessive Pertaining to an allele (see *gene*) which is expressed in the *phenotype* of a diploid organism (see *diploidy*) only when it is in the presence of another identical allele. Compare *dominant*.

recombinant See *gene mapping*.

recombinant DNA *DNA* formed by joining together DNA from two different organisms. In recombinant-DNA technology (a form of *genetic engineering*), the different types of DNA are combined artificially under controlled conditions.

recombination See *crossing-over*.

rectum The terminal part of the *alimentary canal* in vertebrates, which stores *faeces* before they are eliminated from the body via the *anus* or *cloaca*.

recycling The recovery and re-use of products and materials which would otherwise be thrown away, either by reusing items in their original form or by reprocessing them into new materials. See also *nutrient recycling*.

red blood cell or **erythrocyte** The most common type of cell within vertebrate blood: in humans, there are about 4.5—5.5 million red blood cells in each cubic millimetre of blood. Each cell lives for about 120 days and new cells are continuously produced by the red bone marrow at a rate of about 1.5 million per second. All mammalian red blood cells lack a nucleus. Each cell is approximately 8 μm in diameter, 3 μm thick, and shaped like a biconcave disc. This shape increases the ratio of its surface area to volume, improving rates of gaseous exchange by diffusion. Its *cytoplasm* contains *haemoglobin* for oxygen transport from lungs to tissues.

redox reaction A chemical reaction in which *oxidation* and *reduction* occur. Redox reactions occur when electrons are passed from one *electron carrier* to the next in the *electron transport system* and *photophosphorylation*. These redox reactions are *exergonic*; the energy they release is used to manufacture *ATP*.

reducing sugar All *monosaccharides* and some disaccharides (e.g. *lactose* and *maltose*) which have either a free aldehyde (—CHO) or ketone (>=CO) group able to reduce (see *reduction*) copper II ions (Cu^{2+}) to copper I ions (Cu^+). Reducing sugars give a positive result to *Benedict's test*.

reduction A chemical reaction in which an atom or molecule loses an oxygen atom, gains a hydrogen atom, or gains an electron.

reduction division See *meiosis*.

reflex See *reflex action*.

reflex action A rapid, involuntary, unlearned response to a specific stimulus. In vertebrates, the structures involved in a reflex action constitute a *reflex arc*, in which the stimulus is detected by a *receptor cell*, initiating a *nerve impulse* in the sensory *neurone* which transmits it to the *central nervous system* (the *brain* or *spinal cord*). In the simplest type of reflex arc, the sensory neurone stimulates a motor neurone directly and the motor neurone causes the effector organ (muscle or gland) to respond. In other reflex arcs, another neurone (called an association neurone or interneurone) acts as a link between the sensory neurone and motor neurone. During a reflex arc that passes through the spinal cord, messages are also sent to the brain; although the brain does not instigate a spinal reflex, it can modify it. A reflex action modified by learning or experience is called a *conditioned reflex*. In a conditioned reflex, the reflex action is induced by a stimulus (conditional stimulus) other than that which normally produces the unconditioned reflex action.

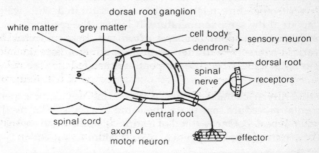

reflex action The reflex arc. The arrows indicate the direction of impulse from receptor to effector.

reflex arc See *reflex action*.

refractory period A period following stimulation, during which a *neurone* or muscle cell becomes inexcitable. During the *absolute refractory period*, immediately after a stimulus, the neurone or muscle cell will not respond to a new stimulus, no matter how intense. During the *relative refractory period*, which follows the absolute refractory period, the neurone or muscle cell will respond only to a stimulus greater than normal.

regeneration The replacement or repair by an organism of a part of its body that has been damaged or removed. In plants, regeneration of whole plants from small portions is common and is used extensively in vegetative reproduction. The ability of animals to regenerate varies from group to group. Starfish and earthworms can regenerate whole individuals from portions of a damaged adult. Crabs sometimes purposely discard a limb when attacked by a predator and regenerate it after several moults. Regeneration in mammals is generally limited to wound healing.

relative humidity A measure of the moisture content of air at a given temperature relative to air fully saturated with water vapour at the same temperature. It reflects the readiness with which water vapour condenses from air. Relative humidity varies inversely with temperature and therefore usually increases at night because temperature then falls.

reliability A characteristic of a measuring device or experimental procedure that yields the same or very similar results when repeated. One method of testing reliability is to compare the results of two independent investigators.

renal Pertaining to the *kidney*.

renal capsule An alternative name for a Bowman's capsule. See *nephron*.

renal tubule See *nephron*.

rennin An *enzyme* secreted by the wall of the *stomach* that coagulates milk proteins. It is especially important to young mammals, because it increases the retention time of milk in the stomach, giving *pepsin* more time to break down its proteins.

reproduction The production of new individuals, ensuring the continued existence of a species beyond the lifetime of an individual. See also *asexual reproduction* and *sexual reproduction*.

reptiles See *Reptilia*.

Reptilia or **reptiles** A class of chordate animals (see *Chordata*), which are scaly-skinned and usually have limbs (snakes and some lizards have secondarily lost their limbs). They have lungs and produce eggs with shells.

rER See *rough endoplasmic reticulum*.

resistance The ability to tolerate an adverse environmental factor such as extreme temperature, a pathogen, or a chemical such as an antibiotic or pesticide. Increased resistance due to a *mutation* is inherited. See also *environmental resistance*.

resolving power The ability of a microscope to differentiate between closely spaced objects; the higher the resolving power of a microscope, the greater its ability to produce clear, detailed, well-defined images at high magnifications. The resolving power is inversely proportional to the wavelength of radiation used. Electron microscopes use electrons of very small wavelength and can achieve resolutions down to 0.2 nm.

respiration The breakdown of organic molecules to release energy in a form (i.e. *ATP*) available to organisms to do biological work. See also *aerobic respiration* and *anaerobic respiration*.

respiratory chain A series of organic molecules on the inner membrane of *mitochondria* that accept hydrogen atoms (protons) and electrons (released from hydrogen) during *aerobic respiration*. The first part of the chain comprises hydrogen carriers which take up hydrogen released from the Krebs cycle; the second part comprises the *electron transport system*. Transfer of hydrogen or electrons from one molecule in the chain to the next constitutes a *redox reaction* which releases energy that can be used to make *ATP*. Oxygen is the final component of the respiratory chain; it combines with hydrogen and electrons to form water.

respiratory quotient or **RQ** The ratio of the volume of carbon dioxide produced in **respiration** to the volume of oxygen taken up in a given time. In animals, the RQ indicates the type of respiration taking place and the substrate being used. An RQ greater than 1.0 indicates *anaerobic respiration*, an RQ of 1.0 or less indicates *aerobic respiration*; aerobic respiration of carbohydrates gives an RQ of 1.0, of proteins 0.9, and of lipids 0.7.

respiratory surface The surface of an organ (e.g. lung or gill) across which *gas exchange* takes place.

respirometer A device for measuring the *respiration* of an organism or tissues. In simple respirometers, a manometer measures the gas volume changes caused by oxygen uptake or carbon dioxide release.

response A change in the activity of a cell or organism as a result of a stimulus.

restriction endonuclease See *restriction enzyme*.

restriction enzyme or **restriction endonuclease** An *enzyme* that will cut *DNA* molecules at specific sequences of base pairs. Restriction enzymes are produced naturally by bacteria as a defence against *bacteriophages* and are used widely in *recombinant-DNA* technology.

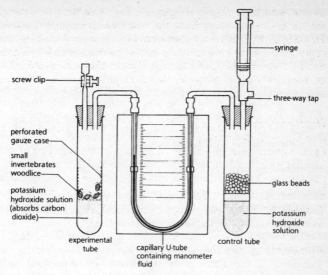

syringe

screw clip

three-way tap

perforated gauze case

small invertebrates woodlice

potassium hydroxide solution (absorbs carbon dioxide)

glass beads

potassium hydroxide solution

experimental tube

capillary U-tube containing manometer fluid

control tube

respirometer

results Data and observations obtained during an experimental investigation; data often take the form of measurements (e.g. of temperature readings, numbers of organisms, etc.). To make sense of results, they should be recorded efficiently and accurately (see *table*), presented in an appropriate and easily understandable format (see *graph*), and analysed using the appropriate method (see *statistics*).

resuscitation The process of restoring to consciousness a person who appears to be dead.

reticulate See *Dicotyledonae*.

reticulum See *rumen*.

retina See *eye*.

retinol See *vitamin A*.

retrovirus See *reverse transcriptase*.

reverse transcriptase An *enzyme* produced by those *viruses* (*retroviruses*, such as the virus responsible for *AIDS*) in which *RNA* is the genetic material. Reverse transcriptase produces a complementary *DNA* copy of the viral RNA, allowing it to combine with the DNA of its host.

rheotaxis See *taxis*.

rhizome A swollen, horizontal, plant stem, usually underground, that bears leaves above and roots below. It usually acts as a perennating organ (see *perennation*).

Rhizopoda A *phylum* within the kingdom *Protoctista*. All members of the Rhizopoda (*rhizopods*) are equipped with pseudopodia (see *pseudopodium*) for locomotion. The best known example is *Amoeba*.

rhodopsin See *rod*.

rib One of the long, curved bones forming the wall of the *thorax* in a tetrapod vertebrate. Ribs are interconnected by the *intercostals* and each rib is attached at one end to a thoracic *vertebra*.

riboflavin See *vitamin B2*.

ribonucleic acid See *RNA*.

ribose A five-carbon (pentose) sugar found in *RNA* and *nucleotides* such as *ATP* and *NAD*.

ribosomal RNA See *RNA*.

ribosome A very small *organelle*, 10—20 nm in diameter, consisting of two subunits, one larger than the other and each made of roughly equal amounts of *protein* and *RNA*, assembled in the *cytoplasm*. Ribosomes are involved in *protein synthesis*, holding interacting molecules together. They may occur freely or be attached to the *endoplasmic reticulum*. Ribosomes of *prokaryotes* are smaller than those lying free within the cytoplasm of *eukaryotes*. Ribosomes in *mitochondria* and *chloroplasts* are the same size as those in prokaryotes, supporting the theory that these organelles may have originated from symbiotic bacteria.

ribulose biphosphate or **ribulose diphosphate** A five-carbon compound that plays a central role in the *Calvin cycle*. It fixes carbon dioxide to produce a six-carbon compound which is converted almost instantly into two molecules of the three-carbon compound *glycerate 3-phosphate* (GP). During the Calvin cycle, sugars are synthesized and ribulose biphosphate regenerated.

ribulose diphosphate See *ribulose biphosphate*.

rickets A *deficiency disease* of young children and other mammals, caused by lack of *vitamin D*, in which bones are malformed because they fail to harden fully.

RNA or **ribonucleic acid** A *nucleic acid* usually consisting of a single chain of nucleotides, with ribose as the sugar, and the bases adenine, cytosine, guanine and uracil. Various forms of RNA are found. These include *messenger RNA* (mRNA), *transfer RNA* (tRNA), and *ribosomal RNA* (rRNA). mRNA consists of a strand of nucleotides, formed by *transcription* of *DNA*, that carries the *genetic code* for the synthesis of a polypeptide chain.

In *eukaryotes*, mRNA moves from the nucleus to the *cytoplasm*, where it acts as a template for the synthesis of polypeptides on *ribosomes* by the process of *translation*. tRNA is a single-stranded, cloverleaf-shaped, polynucleotide chain that carries *amino acids* to the ribosomes for polypeptide synthesis. Each of the 20 or so amino acids is carried by a specific type of tRNA which bears a special region, called the *anticodon*, consisting of three bases complementary with the *codon* on mRNA. The amino acid attaches to one end of the tRNA. The anticodon occurs about halfway along the tRNA and the bases are exposed in such a way that they can temporarily combine by hydrogen bonding to the complementary bases on mRNA. rRNA is a major component of ribosomes. In eukaryotes, it is formed in the nuclear organizing region of the nucleus.

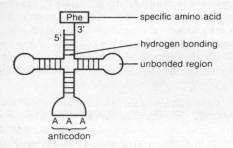

RNA Transfer RNA.

rod One of the *receptor cells* in the retina of the *eye*; it is rod-shaped. Rods contain a light-sensitive, purple pigment (*rhodopsin*) and are responsible for non-colour vision and vision in dim light; they are particularly sensitive to changes in light intensity due to movement in the environment. Rods are absent from the *fovea* and increase in numbers towards the edges of the retina.

root 1. The structure that anchors a plant in the soil and is responsible for the uptake of water and mineral salts. Unlike stems, roots never contain *chlorophyll* and do not bear leaves, flowers or buds. The root tip is covered with a thin layer of cells (the *root cap* or *calyptra*) that protects the growth region (root meristem) just behind it from abrasion as the root grows and pushes its way between soil particles. The cells of the root cap are constantly worn away and replaced by the root meristem. *Root hairs*, growing out of another thin layer of epidermis (the piliferous layer) a short distance behind the root tip, are in intimate contact with the soil. These are fine, thin-walled, tube-shaped structures that greatly increase the surface area of the root for water absorption.
2. See *tooth*.

root cap See *root*.

root hairs See *root*.

root nodule A small swelling on the root of some plants (especially legumes such as pea, gorse and bean) that contains nitrogen-fixing bacteria.

root pressure A force in plant roots that pushes water up the *xylem*. It is observed and measured when a freshly cut root stump continues to exude sap from its xylem vessels. It is thought to be caused by the *active transport* of ions into the xylem sap, thus lowering its water potential. Water is drawn into the xylem from neighbouring cells by *osmosis* and forced up the stem. Root pressure is insufficient to account for *transpiration* in large plants, but it probably plays an important role in moving water up some plants when transpiration rates are low. In the dim light and high humidity of tropical rain forests, for example, root pressure is responsible in some plants for *guttation* (the loss of water droplets from the surface of the plant's leaves through special, permanently open stomata, called *hydathodes*).

fermentation chamber and contains bacteria that digest *cellulose* to *glucose* which is fermented to organic acids. The semi-digested contents of the rumen pass to the second chamber (*reticulum*) where they are formed into balls (the *cud*) and regurgitated to the mouth. After further mastication (chewing the cud, or *rumination*), the cud is swallowed into the third stomach chamber (*omasum*) to be made firmer, and then passes to the fourth chamber (the *abomasum* or true stomach) where protein digestion takes place.

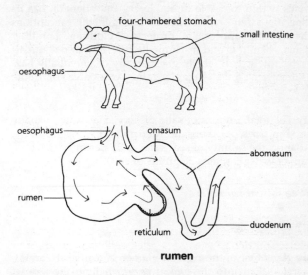

rumen

ruminant See *rumen*.

rumination See *rumen*.

runner A plant stem that runs along the surface of the ground (e.g. in the strawberry plant and creeping buttercup). It has scale leaves, and axillary buds capable of sprouting leaves and growing roots, usually from its outer end, to form a new plant. Unlike a *rhizome*, a runner does not root all along its length and it has no tendency to swell. Its main function is *vegetative reproduction*. See also *stolon*.

safety The condition of being free from danger, essential in all experimental investigations. No investigation can be entirely risk-free, but it is the responsibility of the investigator to minimize risks within tolerable limits. Every investigation has its own specific safety requirements, but general precautions include wearing appropiate protective clothing, handling organisms, chemicals and instruments carefully, and eliminating wastes safely. Full labelling, especially of potentially hazardous items, and a well-organized, tidy work area contribute greatly to safety.

salinity The total amount of salts dissolved in water, usually expressed in parts per thousand (per mille) by weight. In the mid-Atlantic, the salinity of sea water is 35 per mille (i.e. each kilogram of water contains 35 grams of salt).

saliva See *salivary gland*.

salivary amylase See *amylase*.

salivary gland One of the *exocrine glands*, in many vertebrates, that secrete saliva into the *buccal cavity* behind the mouth. Saliva is a watery fluid, consisting mainly of mucus, salts, and (sometimes) salivary amylase, that begins the digestion of starch; the mucus lubricates food before it is swallowed. In some insects (e.g. houseflies) the salivary glands secrete digestive *enzymes* externally on to food. The saliva of blood-sucking organisms usually contains anti-clotting agents.

salt A compound in which the hydrogen of an acid has been replaced partly or wholly by a metal. In common table salt (sodium chloride, NaCl) the hydrogen atom of hydrochloric acid (HCl) has been replaced by sodium. When salts dissolve in water, they form positive and negative ions separated from one another by water molecules.

sample See *sampling*.

sampling The selection of a *sample*; this is any fraction of a whole (e.g. part of a population collected for an investigation, or a portion of a substance to be analysed). See also *random sampling*.

sap Watery liquid contained within the *vascular tissues* of plants or *vacuoles* of plant cells.

saprobiontic nutrition See *biodegradation*.

saprophyte See *saprotroph*.

saprotroph or **saprophyte** An organism which feeds on dead or decaying matter in the form of organic substances in solution. Such organisms play a key role in *biodegradation* and *nutrient recycling*.

sarcomere The functional unit responsible for contraction of striated muscle fibres. Each sarcomere is contained within two *Z-membranes* and is composed mainly of contractile protein filaments (*actin* and *myosin*) which form alternating light and dark bands. The light bands (also called *I bands*) contain mainly the thin actin filaments; the dark bands (also called *A bands*) contain mainly the thick myosin filaments. During contraction, the actin and myosin filaments slide between each other (see *sliding filament theory*), causing the light band to shorten while the dark band stays more or less the same length. See diagram overleaf.

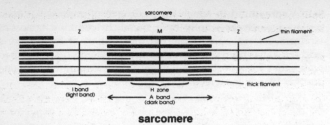

sarcomere

saturated fatty acid A *fatty acid* which has only single covalent bonds between the carbon atoms in the hydrocarbon chain. Saturated fatty acids are easier to pack together and tend to form *fats* rather than *oils*.

scanning electron microscope See *electron microscope*.

scapula The *dorsal* part of the *pectoral* girdle in tetrapods; in humans, the shoulder blade. It is attached to the *clavicle*.

scattergram See *scatter graph*.

scatter graph or **scattergram** A diagram in which each point on a piece of graph paper represents one item in a sample. The position of the point on the graph depends on the magnitude of two variables (e.g. height and weight of an animal); one variable is plotted against the x-axis, the other against the y-axis. A scatter graph is a convenient way of displaying results and gaining an approximate idea of the relationship between the two variables.

Schultze's solution An iodine-containing stain used to indicate the presence and location of *cellulose* and other substances in plant tissue. It stains unlignified cellulose *cell walls* blue or violet, *lignin* yellow, and *starch* blue-black.

Schwann cell A cell that forms the *myelin* sheath encapsulating some *nerve fibre*s, in vertebrates.

scientific method The procedures by which scientific investigations should be conducted. These often involve the following steps: (a) observations are made of a natural phenomenon; (b) a problem is formulated based on questions stemming from the observations; (c) a hypothesis is suggested that provides a possible solution to the problem; (d) predictions are made based on the hypothesis; (e) an experiment incorporating appropriate controls is conducted to test the predictions; (f) the results of the experiment are analysed objectively, if possible, using the appropriate statistics; (g) the hypothesis is accepted, rejected or modified, depending on the results of the experiment.

sclereid See *sclerenchyma*.

sclerenchyma Structural tissue comprising two types of plant cell: elongated fibres and roughly spherical *sclereids* (also called *stone cells*). In both, the *cell wall* is heavily thickened with *lignin*; this has great tensile and compressional strength, so sclerenchyma does not break easily on stretching or buckle easily when a load pushes down on it. Mature sclerenchyma is dead, because once lignification is compete the cells become impermeable to water. Simple pits and tubes of constant thickness pass through the walls of sclerenchyma.

sclerotic layer The tough, fibrous, outer lining of the vertebrate *eye*.

scramble competition See *competition*.

scrotal sac One of two sacs of skin and tissue in most male mammals, that contain and support the testes. The pair of sacs lie outside the body, so the testes are maintained at a slightly lower temperature than the body core, for optimal *sperm* development.

scurvy See *vitamin C*.

sebaceous gland A gland in the skin that secretes oil (*sebum*) through a duct that opens in a hair follicle, from which it is discharged over the skin. Sebum is mildly antiseptic; it lubricates hair follicles and skin and contributes to their waterproofing.

sebum See *sebaceous gland*.

secondary consumer A *carnivore* that eats a herbivore.

secondary growth See *secondary thickening*.

secondary productivity See *productivity*.

secondary sexual characteristics External features, other than *gonads* and associated structures, that distinguish adult male and adult female animals. Their development is controlled by *sex hormones* (e.g. in human males, testosterone induces growth of bodily and facial hair, muscular development, and development of a deep voice; in human females, oestrogens induce breast development, broadening of the pelvis, and deposition of fat around the hips). Human secondary sexual characteristics are permanent, but in some organisms they are temporary (e.g. the red body colour of male sticklebacks occurs only during the mating season).

secondary structure See *protein*.

secondary succession See *succession*.

sectioning

secondary thickening or **secondary growth** The formation of new tissue (secondary *xylem* and *phloem*) in a woody plant by the repeated lateral division of *cambium*. Secondary thickening increases the girth of roots and stems by the addition of successive layers of tissue. In temperate regions and high latitudes, where growth is seasonal, large xylem vessels are formed in the spring and small ones at the end of summer. Consequently, secondary thickening may be visible as annual rings. Secondary thickening occurs mainly in trees and shrubs of *dicots* and *conifers* and provides them with the mechanical support needed to grow tall.

secrete To release a *secretion*.

secretin A polypeptide *hormone* secreted by the small *intestine* in response to acid *chyme* from the *stomach*. It stimulates the release of *bile* from the *liver* and digestive juices from the *pancreas*. Its action was discovered in 1902 and secretin was the first substance to be described as a hormone.

secretion 1. The process by which a useful substance is discharged into the external medium by living cells. In multicellular organisms, secretions are usually produced by glands (see *exocrine gland* and *endocrine gland*).
2. A substance released in this way.

sectioning The cutting of a thin slice of tissue for examination using a microscope. In light microscopy, the tissue is usually supported by freezing or *embedding* it in paraffin wax before it is cut with a metal knife or razor. Hand sectioning is possible for some plant material, but very thin sections are cut using a *microtome*, an instrument which controls the movements of the cutting edge (knife or razor) very precisely. In transmission electron microscopy, extremely thin sections are required, so specimens must be embedded in a very hard substance (e.g. an epoxy resin) and are cut into sections using a glass or diamond knife in a microtome which is often electrically controlled.

sedentary (of an organism) Able to move from one place to another, but with restricted or very slow movement. A sea anemone is a sedentary animal. Compare *sessile*.

seed A reproductive product in flowering plants (*Angiospermatophyta*), formed from the ovule after *double fertilization*. A seed contains a plant *embryo* and food store (either *endosperm* or *cotyledons*) surrounded by a hard, tough, protective coat (the *testa*), which protects the embryo from mechanical damage or attack from microbes.

segmentation See *metameric segmentation*.

segregation See *law of segregation*, *selection* and *natural selection*.

selective reabsorption The process by which specific materials are reabsorbed into the body after they have entered an excretory tubule. In mammals, this is one of the functions of the *kidney*: the *proximal* convoluted tubule of the *nephron* selectively reabsorbs water, glucose and amino acids; the *distal* convoluted tubule selectively reabsorbs water and mineral salts. Selective reabsorption usually takes place against a *concentration gradient* and requires *active transport*.

selectively permeable membrane A membrane through which some particles can pass, but not others. All *cell membranes* are selectively permeable. See *partially permeable membrane*.

self-fertilization See *fertilization*.

semen A fluid containing *spermatozoa* (sperm) and nutrients. In animals with internal *fertilization*, the male discharges semen into the female during copulation. In mammals, sperm are produced in the *testes*, and nutrients in the *prostate gland* surrounding the *urethra*.

semi-permeable membrane See *partially permeable membrane*.

semicircular canals Fluid-filled tubes in the inner *ear* concerned with the perception of body position and balance. There are usually three tubes, one in each plane and positioned at right angles to each other. Movement of the head causes fluid in the tubes to stimulate receptor cells at the end of each. Information about head movement is conveyed to the brain by the auditory nerve, so balance and posture can be maintained.

semilunar valves See *heart*.

sense organ A part of the body of an animal containing a concentration of *receptor cells* which respond to specific internal or external stimuli. Examples are the *eye* and the *ear*.

sensitivity See *irritability*.

sensory cell See *receptor cell*.

sensory neurone A *neurone* that carries information from a *receptor cell* to the *central nervous system*. In vertebrates, sensory neurones enter the *spinal cord* via the dorsal surface and cell bodies occur in ganglia (see *ganglion*) outside the spinal cord.

sepal See *flower*.

sepsis The destruction of tissue by disease-causing bacteria or their toxins.

septum See ***heart***.

seral changes See ***succession***.

sere See ***succession***.

serum Watery fluid of similar composition to blood plasma, but lacking ***fibrinogen*** and clotting factors.

sessile (of an organism, usually an animal) Incapable of moving from one place to another. A barnacle is a sessile animal.

sewage Liquid waste that is removed through sewers, which are pipes carrying both industrial (e.g. from abattoirs, factories and hospitals) and domestic waste (e.g. human faeces and urine, and waste water from sinks). The typical composition of sewage is 99.9% water and only 0.1% solids. Of the solids, approximately 70% are organic and 30% inorganic (metals, salts, and grit). Raw sewage also contains a large number of living organisms, including oxygen-consuming bacteria (see ***biochemical oxygen demand***), some of which can cause disease (i.e. are pathogenic). Sewage treatment includes the removal of sludge by sedimentation in tanks, screening to remove large particles, and aeration to encourage ***biodegradation*** by aerobic bacteria and other organisms. Water intended for human consumption is usually filtered and sterilized.

sex See ***sex determination***.

sex chromosome A ***chromosome*** that is linked with the sex of its bearer. See ***sex determination***.

sex determination The mechanism by which it is determined whether an individual is male or female. In some organisms this is determined entirely genetically. In mammals and many other animals (e.g. *Drosophila*), it is determined by the presence or absence of a *Y chromosome*. Males have an *X chromosome* and Y chromosome in their diploid body cells and are heterogametic (i.e. they produce two dissimilar types of *gametes*, one with an X chromosome and the other with a Y chromosome). Females are homogametic (i.e. their diploid cells usually contain two X chromosomes and all gametes carry an X chromosome). In birds, the female is heterogametic (XY) and the males homogametic (XX). In other organisms, sex may be greatly influenced by environmental factors. Some (e.g. certain molluscs) change sex with age, others when environmental conditions change (e.g. in some fish, sex is determined by environmental temperature).

sex hormone A *hormone* responsible for development of the reproductive organs or *secondary sexual characteristics*. Human sex hormones include *oestrogen* and *progesterone* in females, and *testosterone* in males.

sex linkage The tendency of certain inherited characteristics to occur more frequently in one sex than the other. This usually results from *genes* (other than those involved in determining *primary* or *secondary sexual characteristics*) carried on the X chromosome. Recessive *phenotypes* tend to predominate in the heterogametic sex, because there is no possibility of the Y chromosome carrying a *dominant* allele to mask a *recessive* allele on the X chromosome. For example, in humans, *haemophilia* and *colour blindness* are sex-linked characteristics much commoner in males than in females.

sexual behaviour Behaviour associated with sexual reproduction, including behaviour preceding, during and after mating. See also *courtship*.

sexual intercourse The term by which human *copulation* is usually known.

sexual reproduction Reproduction that involves the fusion of haploid nuclei, usually *gametes*, to produce a diploid *zygote*. *Meiosis* occurs at some stage in the life cycle of sexually-reproducing organisms. Unlike *asexual reproduction*, sexual reproduction increases variability within a species. In many sexually-reproducing organisms, elaborate mechanisms have evolved to bring male and female gametes together.

sexually transmitted disease Any disease that can be transmitted during *sexual intercourse* from an infected person to another person. Examples are *AIDS* and *gonorrhoea*. The infected person may be a carrier of the disease, exhibiting no symptoms and totally unaware that he or she is infected. Sexually transmitted diseases tend to be spread by people who have more than one sexual partner and who have unprotected sex (i.e. sexual intercourse without wearing a condom).

shin bone See *tibia*.

sickle cell anaemia A hereditary blood disease in humans caused by a gene *mutation* that results in a change in an *amino acid* in one of the chains of *haemoglobin*. Sufferers have two *recessive* alleles per cell. Some of their *red blood cells* become sickle-shaped, reducing the ability of blood to carry oxygen and increasing the tendency of blood cells to clump together. Those with only one mutant allele in each cell are said to have *sickle cell trait*. They usually suffer few symptoms and tend to have a greater resistance to malaria. This may explain why the sickle cell allele is usually found in people who live in, or originated in, malarial areas of Africa.

sickle cell trait See *sickle cell anaemia*.

short-sightedness See *myopia*.

shoulder blade See *scapula*.

sieve tube See *phloem*.

sieve tube element See *phloem*.

simple sugar See *monosaccharide*.

single circulatory system See *circulatory system*.

sinoatrial node See *pacemaker*.

size Dimensions or extent; the size of an organism may be based on its linear dimensions, girth, volume, dry mass or wet mass. See also *growth*.

skeletal muscle *Muscle* that is attached to a bony or cartilaginous *skeleton*. The contractile tissue of skeletal muscle comprises fibrils with alternating dark and light bands. Each muscle fibre is multinucleated and made up of *sarcomeres*. All skeletal muscles involved in voluntary movements are *striated muscles*.

skeleton Any structure that supports the body and maintains its shape. Cells have a subcellular *cytoskeleton*. Multicellular animals have one of three types of skeleton: an internal, bony or cartilaginous *endoskeleton*; a hard, external covering that acts as an *exoskeleton*; or fluid-filled compartments which act as a *hydrostatic skeleton*.

skin The outer covering of an animal, external to the main musculature, consisting of an epidermis and dermis; the surface is often covered by scales, feathers, or hair. The skin acts as a protective layer, forming the first line of defence against invasion by bacteria and other foreign bodies, and containing pigments that protect underlying structures from harmful solar radiation. In reptiles, birds and mammals, the skin is waterproof. In mammals, the skin contains *sense organs* responsive to touch, pressure, pain, heat and cold, enabling them to be aware of their surroundings. Fats situated just below the skin (*subcutaneous fats*) provide thermal insulation and also act as an energy store. Other features of mammalian skin which contribute to *thermoregulation* include the ability to vary the blood supply to the skin surface, variations in secretion of sweat from *sweat glands*, shivering, and changes in hair positions.

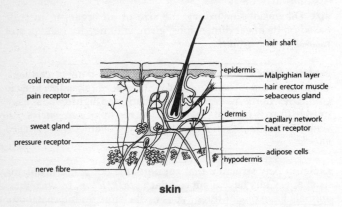

skin

skull The *skeleton* of the head in vertebrates. See also *cranium*.

sliding filament theory The theory that when *muscle* contracts, individual filaments of *actin* and *myosin* do not contract but slide between one another, thus shortening the *sarcomere*. During contraction, the globular heads of the thick myosin filaments act as cross-bridges which attach and reattach to the thin actin filaments in a ratchet-like manner, causing the sliding action. Attachment of the cross bridges requires energy (*ATP*) and depends on the presence of calcium ions released over a sarcomere when it is stimulated by an excitatory *nerve impulse*.

slime layer See *cell capsule*.

small intestine The *anterior* region of the vertebrate *intestine*. In humans, it consists of a *duodenum* (about 30 cm long) which receives food from the stomach, a small portion called the *jejunum*, and the *ileum* (about 7 m long) from which food passes to the large intestine.

smell or **olfaction** The sense by which animals detect airborne chemicals. In mammals, the nose is the organ of smell (*olfactory organ*). Its cavity (nasal cavity) contains receptor cells for smell (olfactory hairs); cilia on the hairs act as receptor sites to which chemicals can bind, depolarizing the hair cell and causing a *nerve impulse* to be sent to the olfactory centres of the brain. The sense of smell is very similar to that of taste, and in aquatic environments there is really no distinction.

smog 1. Originally, a mixture of smoke and fog trapped beneath a temperature inversion and experienced in certain industrial cities (e.g. London) where coal was the principal domestic and industrial fuel.
2. (photochemical smog) A form of air pollution which arises from the action of strong sunlight on nitrogen oxides and hydrocarbons emitted from motor vehicles in humid conditions when the ingredients are trapped beneath a temperature inversion. It occurs in many cities, especially in low latitudes; examples are Los Angeles, Mexico, Athens.

smoking The inhalation of products released from the combustion of tobacco, including nicotine and carbon monoxide. Tobacco products carry a warning stating that smoking can damage your health.

smooth endoplasmic reticulum See *endoplasmic reticulum*.

smooth muscle or **involuntary muscle** *Muscle* that consists of spindle-shaped cells with no obvious striations. Smooth muscle lines the walls of hollow organs (e.g. the *stomach* and *blood vessels*). It is particularly well adapted to producing long, slow contractions (e.g. *peristalsis*) which are not under voluntary control.

soil Unconsolidated material that forms the surface of land and in which plants gain support, protection, water and nutrients. Soil is produced by the weathering of rock into inorganic particles with the addition of organic matter, air and water. Factors affecting the community of organisms within the soil are known as *edaphic factors*.

soluble (of a substance) Capable of being dissolved in a *solvent* (usually water).

solute A substance that is dissolved in a *solvent*.

solute potential or ψS A change in water potential due to the presence of solute molecules. Solute molecules always reduce water potential, therefore solute potential is always negative. Solute potential was formerly referred to as *osmotic potential*.

solution A homogeneous mixture (i.e. one that is uniform throughout) of two or more substances in which the molecules or atoms are completely dispersed. Compare *colloid*.

solvency The *solvent* properties of a substance.

solvent A component forming the greater part of a *solution*, or whose physical state (usually liquid) is the same as that of the solution. Water is the most important solvent in organisms. So many substances dissolve in it, it is sometimes called the 'universal solvent'.

Spearmann rank correlation coefficient A measure of the strength of the association between pairs of measurements of a sample (sample size 7—30). For each member of the sample, the rank orders of the two measurements are arranged side by side in two columns so they can be compared statistically. In a perfect positive correlation (i.e. correlation coefficient 1.0), the specimen that ranked 1 for one measurement also ranks 1 for the other measurement, and so on, so there are no differences in the ranks. The procedure cannot be used on pairs of measurements which, when plotted on a *scattergram*, form a U- or inverted U-shaped curve.

$$r_s = 1 - [6\Sigma D^2/n(n^2 - 1)],$$

where r_s is Spearmann's rank correlation coefficient, D is the difference in rank between a pair of measurements, Σ is the sum, and n is the sample size.

speciation The process by which two or more new species evolve from one original species; Darwin called it the 'mystery of mysteries'. The original species must first split into two or more separate populations. *Allopatric speciation* occurs when two or more populations of the original species become physically separated by geographical barriers that prevent interbreeding. *Sympatric speciation* occurs when two populations living in the same area are isolated by some mechanism other than geographical isolation (see *isolating mechanisms*). Once isolated, *mutation* and *natural selection* act independently within the two populations, which develop into two distinct species. Parapatric speciation occurs in partially isolated populations under strong selection pressure, despite a small flow of genes between the populations.

species A group of similar organisms whose members can, at least potentially, interbreed to produce fertile offspring. See also *binomial nomenclature*.

species diversity A measure of the biological richness of a community or area, taking into consideration both the *species richness* (number of species present) and *species equitability* (the relative abundance of different species). There are several mathematical formulae for expressing the species diversity as a single figure (a species diversity index), but no general agreement about which is best. A commonly used index is that of Simpson, which uses the following formula:

$$D = N(N - 1)/\Sigma_n(n - 1),$$

where D is the diversity index, N is the total number of organisms, n is the number of individuals of each species, and Σ is the sum.

species equitability See *species diversity*.

species richness See *species diversity*.

specific heat capacity The quantity of heat required to raise the temperature of a unit mass of a substance by one degree Celsius.

specificity The precision with which an action is directed to a particular target, affecting nothing else. Pesticides, for example, are evaluated according to their specificity. It is a characteristic feature of *enzymes*, which catalyse specific types of reaction involving only one substrate or group of substrates (group specificity). Enzymes are classified according to their substrate and substrate specificity. Thus, the technically correct name for catalase is hydrogen peroxide—hydrogen peroxide oxidoreductase.

specific name See *binomial nomenclature*.

sperm See *spermatozoon*.

spermatozoon (*pl.* **spermatazoa) or sperm** The mature, motile *gamete* produced by the **testis** of animals. It consists of a head, containing a haploid nucleus and an *acrosome* with hydrolytic *enzymes* to penetrate an egg cell during *fertilization*; a middle region, containing *mitochondria* which provide energy for movement; and a tail section, containing *microtubules* similar to those of cilia. Locomotion is by movements of the tail section.

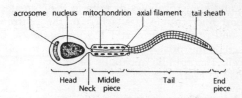

acrosome   nucleus   mitochondrion   axial filament   tail sheath

Head   Middle   Tail   End
Neck   piece   piece

spermatozoon Length about 6μm.

sphincter Circular *muscle* that closes or contracts an opening. The *pyloric sphincter* is an example.

spinal column See *vertebral column*.

spinal cord Part of the *central nervous system* of vertebrates, extending backwards from the *brain* as a uniform tube enclosed and protected by the vertebrae. It contains an inner area of grey matter and an outer area of white matter, a central cavity containing cerebrospinal fluid, and an outer layer of membranes (the meninges). Sensory *neurones* enter the dorsal part (dorsal root) of the spinal cord and motor neurones leave from the ventral region (ventral root). Both sets of neurones combine to form pairs of mixed nerves (each called a spinal nerve) which leave the cord in each segment of the body.

spinal nerve See *spinal cord*.

spinal reflex A reflex arc (see *reflex action*) which passes through the *spinal cord*.

spindle apparatus The structure formed during nuclear division to which *homologous chromosomes* or sister *chromatids* attach prior to separating. It consists of long *microtubules* radiating from end to end of the cell and shorter microtubules extending from one end of the cell to the equator (the region where spindle microtubules diverge to their greatest extent). Each short microtubule is attached to the *centromere* of a chromosome or sister chromatid. During *anaphase*, the microtubules guide the movement of chromosomes to opposite ends of the cell. The spindle apparatus of animal cells is formed from a pair of *centrioles*.

spiracle 1. (in insects) The opening of a *trachea* to the atmosphere. Spiracles often possess valves which can close to minimize water loss.
2. The opening (a modified gill slit) behind the eye of a cartilaginous fish (e.g. a dogfish), through which water passes for *gas exchange*.

spirilli See *spirillum*.

spirillum (*pl.* **spirilli**) A rigid, spiral-shaped bacterium.

spleen An abdominal organ in most vertebrates, that produces *white blood cells* and destroys worn out *red blood cells*. In humans, it is able to act as a blood reservoir, storing up to 30% of all blood cells. It is very delicate and is easily damaged during an accident.

spongy mesophyll See *mesophyll*.

spontaneous generation The idea, widely held prior to the 19th century, that organisms can arise spontaneously from non-living material. It was shown by Pasteur not to occur under present-day conditions. He conducted a series of experiments using swan-necked flasks of sterilized hay and water. Although air could enter the flasks, bacteria were trapped in the U-bend and no living organisms materialized.

spore A small, usually single-celled, reproductive body from which a new organism arises when conditions are suitable for germination. Plant spores are haploid (see *haploidy*). Spores contain no *embryo*; this distinguishes them from *seeds*.

sporophyte The diploid generation that produces haploid *spores* by *meiosis* in plants that exhibit an *alternation of generations*.

stabilizing selection See *natural selection*.

staining (in microscopy) The process by which different parts of a specimen are made easier to observe. A stain (a chemical which can bind to a cellular component) heightens the contrast between cell or tissue components, making some parts opaque or coloured and leaving others transparent or colourless. Iodine solution, for example, stains starch blue-black. Stains (e.g. osmium tetroxide) used in electron microscopy are opaque to electrons.

stamen See *flower*.

standard deviation A statistical index of the variation of a set of values from the arithmetical *mean*. It is the square root of the variance. Therefore

$$S = \sqrt{[\sum(x - \bar{x})^2/n - 1]},$$

where S is the standard deviation, $\sum$ is the sum, x is an individual measurement, n is the number of measurements in the sample, and $\bar{x}$ is the mean of the sample.

standing crop See *biomass*.

standing stock See *biomass*.

stapes or **stirrup** A small, stirrup-shaped bone in the middle **ear** that transmits vibrations from the eardrum to the inner ear. See *ear ossicle*.

starch A polysaccharide, $(C_6H_{10}O_5)_x$, formed by *condensation* of alpha glucose molecules. Natural starches contain a mixture of two components: straight-chained *amylose* and branch-chained *amylopectin*. Each contains alpha glucose monomers linked together by 1,4 *glycosidic bonds*. The enzyme *amylase* accelerates the breakdown of starch by *hydrolysis*, first to dextrins (short chains of glucose monomers) and then to *maltose*. Starch molecules are synthesized during the light-independent stage in chloroplasts where they are laid down in a series of concentric rings to form starch grains. Starch is relatively insoluble and has little effect on the solute potential of cells, consequently it also has little effect on the osmotic movement of water in a plant. Starch is the main storage *carbohydrate* in most plants, occurring especially in roots, tubers, seeds and fruits. Starch is a major energy source for animals. Starch combines with *iodine* to form a purple—black coloured polyiodide complex.

starch

starch grain See *starch granule*.

starch granule or **starch grain** An accumulation of *starch* which can be seen, with the aid of a microscope, in chloroplasts and plant storage cells (e.g. in the potato *tuber*). Each starch grain appears to consist of concentric layers of starch, the size and shape of which is usually characteristic for a given plant species.

starch test or **iodine test** A simple test to detect the presence of *starch*. A few drops of iodine or potassium iodide solution are added to the test substance, in solution or as a solid. A blue—black coloration indicates the presence of starch which forms a polyiodide complex with the iodine.

statistics The branch of mathematics concerned with the collection, classification and interpretation of quantitative data and the application of probability theory to analyse data. Statistical tests or methods (sometimes abbreviated to statistics) are used in scientific investigations to analyse sets of results so they can be interpreted objectively and reliable conclusions made. The statistical test must be suitable for the data being analysed and the problem being investigated. See *chi-squared test*, *Mann—Whitney U test*, *Spearmann rank correlation coefficient*, *standard deviation*, and *t-test*.

stem The main body axis of a *tracheophyte* (vascular plant), bearing leaves or scale leaves, buds and flowers. Most stems are erect and above ground. Their main task is to hold the leaves in a suitable position for *photosynthesis* and the flowers in a suitable position for pollination. Some stems function as underground perennating organs (e.g. *rhizomes* and *tubers*); others, (e.g. the strawberry runner) are used for *vegetative reproduction*.

sterile 1. Unable to produce offspring.
2. Free from living microorganisms. In microbiology, this applies especially to culture media and equipment.

sterilization 1. The process by which an organism is prevented from reproducing (e.g. by the removal of the *gonads*).
2. Rendering an object sterile by removing all microorganisms (chemically, by filtration, or by exposure to heat or radiation).

sternum or **breast bone** The bone in the middle of the ventral region of the chest in tetrapods, to which the ventral end of most *ribs* are attached. The sternum is attached anteriorly to the pectoral girdle. In flying birds, it is a large, keel-shaped bone to which the flight muscles of the wings are attached.

stigma See *flower*.

stimulus A detectable change in the external environment or internal state of an organism that is capable of causing the organism or any of its parts to react or change, but which does not provide the energy for the response.

stimulus detection The process by which a *stimulus* causes a *receptor cell* to send a *nerve impulse* along a sensory *neurone*.

stirrup See *stapes*.

stolon 1. (in plants) Sometimes a synonym for *runner*, but more usually a long aerial stem (e.g. of blackberry) that grows laterally, eventually bends over under its own weight, and gives rise to a new plant when the bud at its apex touches the ground.

2. (in fungi) An aerial *hypha* (i.e. one that grows above the substrate).

stoma (*pl*. **stomata**) One of the numerous small pores in leaves (and sometimes stems) of terrestrial plants, through which gas and water vapour are exchanged. The size of the stomatal aperture is regulated by two *guard cells* which surround each stoma. The *cellulose* wall of the part of the guard cell lining the stoma is thicker than other parts and when the guard cells become turgid they bend, opening the stoma; when the guard cells become flaccid, the stoma closes.

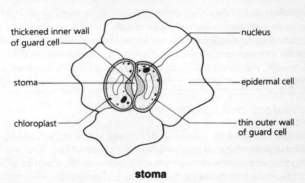

thickened inner wall of guard cell

nucleus

stoma

epidermal cell

chloroplast

thin outer wall of guard cell

stoma

stomach An organ in the *alimentary canal* of vertebrates, consisting of a large chamber with muscular walls, into which food is passed from the *oesophagus*. In carnivores, food may be stored in the stomach for long periods, so they need not eat continuously; in herbivores, the stomach may be adapted to deal with a diet rich in cellulose (see *rumen*). The walls of the stomach secrete gastric juices containing hydrochloric acid and proteases which begin the digestion of protein. Partially digested food (chyme) is discharged to the *duodenum* through the *pyloric sphincter*.

stomata See *stoma*.

stone cell See *sclerenchyma*.

storage material Any substance that is accumulated by a cell for future use. The major storage *carbohydrates* are *glycogen* in animals cells and *starch* in plant cells. Other storage materials include *sucrose* (e.g. in onion bulbs), and *lipids* (e.g. in *adipose tissue*). The main carbohydrate store in grasses is a polymer of *fructose* called inulin.

stratification Subjecting certain seeds to a period of cold in order to encourage them to germinate. In nature, stratification occurs during winter in seeds covered in soil, leaves etc.; and it can be carried out artificially by placing seeds between layers of moist substrate (e.g. sand, moss, or peat) kept at low temperatures.

striated muscle See *skeletal muscle*.

stroma The gel-like matrix of a *chloroplast*, in which the **light-independent stage** of *photosynthesis* takes place. The stroma contains *enzymes* required for the *Calvin cycle*, circular DNA, and *ribosomes* similar to those found in bacteria.

strong acid See *acid*.

structure 1. (in biology) The way different parts of a system (e.g. ecosystem, organism, organ, cell or biological molecule) are arranged.

2. A physical object constructed from component parts.

style See *flower*.

subcellular fractionation or **differential centrifugation** A technique for separating the individual components of cells. The cells are placed in a solution and broken up by grinding in a special blender (called a tissue homogenizer), or by ultrasonic waves. In order to minimize damage and slow down reactions, the solution is kept very cold and has the same salt concentration and pH as the cells. The solution produced by the grinding is a homogeneous mixture of cell membranes and organelles (called a *homogenate*). The components of the homogenate are separated by *ultracentrifugation*. Nuclei and chloroplasts sediment in 10 minutes at 100—500 g; mitochondria and lysosomes in 20 minutes at 10 000—20 000 g; and rough endoplasmic reticulum and ribosomes in 60 minutes at 100 000 g.

submucosa See *duodenum*.

substitution See *mutation*.

substrate 1. A reactant in a chemical reaction catalysed by an *enzyme*.
2. A surface on or in which an organism lives.

subtilisin An *enzyme*, secreted by the bacterium *Bacillus subtilis*, that catalyses the breakdown of *proteins*.

succession Progressive, long-term changes in the composition of communities brought about by organisms themselves. For example, bare soil is first colonized by pioneer species, often short-lived and with a high reproductive potential (see *r-species*). Over a period of time, the pioneer community modifies the environment (e.g. by adding nutrients to the soil), enabling new species to colonize and making it less suitable for the pioneer species. This process continues in stages, with one community replacing another, until the final, relatively stable stage (the *climax*) is reached. *Primary succession* (like that on bare soil) occurs in a previously uninhabited environment; *secondary succession* (e.g. on a ploughed field) takes place in an environment where life is already present but has been altered (for example, by natural disaster or human interference). Each stage in succession is called a *sere* and the changes are sometimes called *seral changes*. Succession may take many years to complete. In England, for example, succession from ploughed field to oak woodland may take 150 years or more.

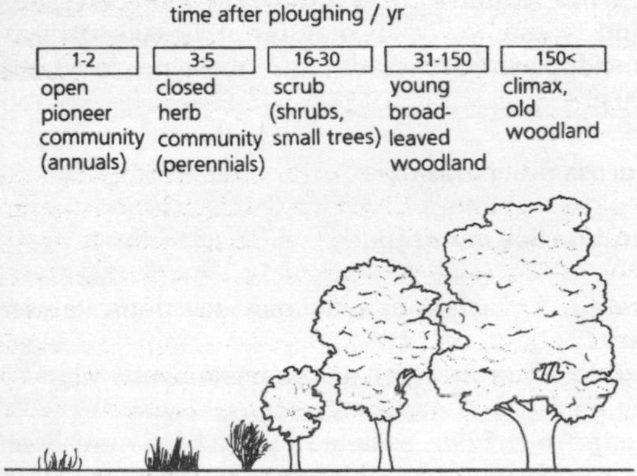

time after ploughing / yr

1-2	3-5	16-30	31-150	150<
open pioneer community (annuals)	closed herb community (perennials)	scrub (shrubs, small trees)	young broad-leaved woodland	climax, old woodland

succession

sucrase or **invertase** A digestive *enzyme*, secreted by the *duodenum*, that catalyses the breakdown of *sucrose* into *fructose* and *glucose*. It also helps to breakdown *maltose* into *glucose*.

sucrose A *disaccharide* formed by the combination of *fructose* and *glucose*; a non-reducing sugar, formula $C_{11}H_{22}O_{11}$. Sucrose is an important energy storage product in some plants (e.g. sugar cane and sugar beet from which table sugar is extracted). It is also the principal form in which *carbohydrate* is transported in the phloem of plants.

Sudan III A red dye that is retained by *lipids*, which it is therefore used to identify and stain. If Sudan III is added to a mixture of oil and water and shaken, the oil becomes red-stained and separates out on the surface of the water which remains uncoloured.

sugar A member of a group of simple *carbohydrates* which share the characteristics of being sweet-tasting, crystalline, and soluble in water. Sugars are classified chemically as *monosaccharides* and *disaccharides*. Table sugar (see *sucrose*) is a disaccharide made from *glucose* and *fructose*.

sulphur dioxide A product of the combustion of some *fossil fuels* which contributes to atmospheric pollution and has been implicated as a causal agent of *acid rain*. It is released naturally during volcanic activity. Some species of lichens are particularly sensitive to sulphur dioxide and are used as indicator species. Sulphites, salts derived from sulphur dioxide, are used as preservatives to prevent oxidation and mould growth in many processed foods and beverages, including wines.

surface area : volume ratio The surface area of an object divided by its volume. As a three-dimensional structure with a regular shape (e.g. a cube) becomes larger, the ratio of its surface area to its volume decreases. This also applies to regularly shaped biological structures. Some activities are directly related to surface area (e.g. *diffusion*, heat loss and water loss increase with increasing surface area) and others are related to volume (e.g. heat production by metabolism and mass increase with increasing volume). This has important biological consequences — for example, small organisms, such as unicells, gain adequate oxygen by diffusion across the whole of their body surface, but large multicellular organisms require special, highly folded surfaces for gas exchange.

surface tension The force that causes the surface of a liquid to contract so that it occupies the least possible surface area. It results from *cohesion* between molecules at the surface of the liquid. See *water*.

surfactant See *detergent*.

survival Staying alive. According to Darwin's theory of evolution by *natural selection*, individuals in a species vary and the ones most likely to survive are those best adapted to their particular environment. This is sometimes called 'survival of the fittest', but it does not imply that individuals necessarily struggle against each other to survive.

suspensory ligaments See *eye*.

swallowing or **deglutition** The *reflex action* by which food is passed from the *buccal cavity* and down the *oesophagus* to the *stomach*. The presence of food in the *pharynx* initiates swallowing, during which the epiglottis closes, blocking the opening to the trachea, and muscles in the oesophagus contract to produce peristaltic waves (see *peristalsis*) which push the food into the stomach.

sweat gland A gland in the skin of many mammals, which secretes sweat through a duct to the skin surface. Sweat is a watery fluid containing salts and *urea*. It plays a small part in excretion, but a large part in temperature regulation, because the evaporation of sweat causes heat to be lost from the skin surface, cooling the body. The distribution of sweat glands varies from species to species. In humans, they are numerous (each individual has approximately 2.5 million) and occur over most of the body surface.

symbiosis The situation in which two or more organisms of different species live together in close association. In this sense the term includes *commensalism*, *mutualism* and parasitism (see *parasite*). However, the term symbiosis is often used as a synonym of mutualism.

sympathetic nervous system See *nervous system*.

sympatric speciation See *speciation*.

synapse A microscopic gap of about 40 nm, between the axon of one *neurone* and the dendrite of another. *Nerve impulses* arriving at the bulbous end of an axon (called a *synaptic knob*), cause the release of a chemical (a *neurotransmitter*) which diffuses across the synapse and attaches to receptor sites on the membrane of the next neurone, changing the potential difference across the membrane. If enough neurotransmitter is released by excitatory impulses, a nerve impulse is transmitted along the next neurone.

synaptic knob See *synapse*.

synecology See *ecology*.

synovial joint See *synovial membrane*.

synovial membrane Loose *connective tissue*, lining the inside of the capsule of a freely moving joint (*synovial joint*), which secretes a fluid (synovial fluid) that is contained within the joint cavity, where it lubricates and nourishes the joint and acts as a cushion preventing the joint surfaces from rubbing against one another.

synthesis The formation of complex substances from simpler ones. In organisms, syntheses are *anabolic reactions* which require energy.

system A distinct entity consisting of an organized set of inter-related and interactive parts, such that a change in one part may affect the entity as a whole. Biological systems include *cells*, *organs*, *organisms* and *ecosystems*.

systematics The whole study of biological classification and the diversity of living organisms; the term is sometimes used as a synonym for *taxonomy*.

systemic circulation That part of a double *circulatory system* in which blood passes from the left ventricle to all parts of the body except the lungs.

systole See *cardiac cycle*.

systolic pressure See *blood pressure*.

table or **tabulation** One of the simplest ways of presenting data, in which values for two or more related variables are displayed in columns. Tabulation of data is the first step in recording information. For each variable, the values given in a table should be pure numbers. Each column in which these are displayed should be headed with a physical quantity, the appropriate SI unit for the variable. The name of the variable should be separated from the quantity by a solidus (/); i.e. a rate of 5 grams per second might appear in the text as 5 g s^{-1}, but in a table should be presented as rate/g s^{-1}, or rate/g per s.

root tip See *root*.

rough endoplasmic reticulum or **rER** A system of parallel *cell membranes* enclosing fluid-filled channels (*cisternae*). The cell membranes are covered by *ribosomes*. These synthesize polypeptides which are secreted into the cisternae where they are combined into proteins or carried to the *Golgi body* for further processing and packaging.

roughage Indigestible plant material, consisting mainly of *cellulose*, that forms an important part of the human diet. Roughage absorbs water and provides most of the bulk against which the muscles of the *intestine* can work, thus improving the movement of food in the gut and reducing the risk of constipation. Some roughage is broken down by bacteria in the large intestine, releasing substances that may have beneficial effects on humans (by, for example, reducing *cholesterol* levels).

round window or **fenestra rotunda** A small, round, membraneous area that connects the inner to the middle *ear*.

RQ See *respiratory quotient*.

rRNA See *RNA*.

r-species Species that have the potential to reproduce rapidly; they are usually opportunistic pioneer species of unstable habitats which can take advantage of brief periods when environmental conditions are optimal. Their population growth tends to show *density-independence* and they usually have *J-shaped growth curves* of the 'boom-and-bust' variety. They tend to be small and short-lived, and poor competitors.

rumen The first chamber of a four-chambered stomach, derived from the lower part of the *oesophagus*, in herbivorous mammals called *ruminants* (e.g. cows and sheep). It acts as a taxis

In a table with two or more columns, the first column should, if possible, contain data for the independent variable (i.e. the variable whose values are chosen by the investigator) and the second and subsequent columns values for the dependent variable. Each table should have an informative title. For tables displaying qualitative information, any abbreviations or symbols used should be explained clearly in notes (a legend) at the bottom of the table.

tabulation See *table*.

tail The rearmost part of an animal. Vertebrates have a *post-anal tail*.

tarsus The ankle bone in tetrapods.

taste or **gustation** The ability of terrestrial animals to detect specific environmental chemicals in a solution. (In an aquatic environment there is no distinction between taste and *smell*.) In humans, detection occurs in special organs (taste buds) in the *buccal cavity*, especially on the tongue. Four types of *receptor cells*, located in different regions of the tongue, allow four types of taste to be distinguished (salt, sweet, bitter and sour).

taxa See *taxon*.

taxis A change in the direction of movement of a whole organism or *cell* in response to an external *stimulus*. The type of stimulus is specified in the name given to the taxis (e.g. *chemotaxis* is induced by particular chemical substances, *phototaxis* by changes in light intensity, *rheotaxis* by water or air currents, *geotaxis* in response to gravity). Movement towards the stimulus is positive (+) taxis; movement away from it negative (−) taxis. For example, woodlice will move away from light (negative phototaxis) and male moths will move towards the scent of a female moth (positive chemotaxis).

taxon (*pl*. **taxa**) or **taxonomic group** An assemblage of organisms that share some basic features. Each taxon has a particular status or *rank* (e.g. phylum) which it shares with other taxa, and the taxon is given an individual name (e.g. Arthropoda). There are seven levels of taxon in most biological classifications: kingdom, phylum (in plants, often called a division), class, order, family, genus, species.

taxonomic group See *taxon*.

taxonomy The study of the principles, practice, and rules for the *classification* of living and extinct organisms.

T cell See *T-lymphocyte*.

teleost See *Osteichthyes*.

telophase The stage in nuclear division when two sets of *chromosomes* come together at opposite ends of a cell to form two daughter nuclei. During telophase, the *spindle apparatus* disappears, the nucleoli and the nuclear membrane reappear, and the chromosomes become dispersed as *chromatin*.

temperature A measure of hotness; the property determining the rate at which heat will be transferred between two bodies in direct contact (heat flows from regions of high temperature to regions of low temperature). See also *thermoregulation*.

temperature coefficient See Q_{10}.

tendon A band of *connective tissue*, containing mainly *collagen* fibres, that connects a muscle to a bone. Tendons contribute to the effectiveness of muscles by concentrating the pull of the muscle on a small area of bone.

terrestrial habitat A *habitat* on land as opposed to one in water (aquatic habitat) or in the air. Compared with an aquatic habitat, the advantages of a terrestrial habitat include less fluid resistance for locomotion, high light intensity for photosynthesis, and plentiful supply of oxygen; the disadvantages include greater extremes of temperature, mechanical support provided only by soil, not water, and high risk of water losses (this makes transfer of *gametes* particularly difficult).

tertiary consumer A carnivore that preys on other carnivores.

tertiary structure The three-dimensional arrangement of a polypeptide chain, which is usually maintained by *hydrogen bonds* and disulphide bonds (between two sulphur atoms in adjacent molecules of the *amino acid* cysteine). The tertiary structure is vital to the action of *enzymes* and molecule-carriers in *cell membranes*. *Denaturation* by extremes of *pH* and temperature results in disturbance of the tertiary structure. See also *proteins*.

testa See *seed*.

test cross (in genetics) Sexual reproduction between an organism of unknown *genotype* with another organism which possesses the homozygous *recessive* genotype for a particular characteristic. Test crosses are performed to determine whether an organism with a dominant *phenotype* has two *dominant* alleles (i.e. is homozygous) or has a dominant and recessive allele (i.e. is heterozygous). When the individual of known genotype is a parent, the procedure is called a *back cross*.

testis The main reproductive organ in male animals, which produces spermatozoa. In mammals, there are two testes, usually contained within a scrotal sac, that secrete *hormones* (see *testosterone*) in addition to producing *sperm*.

testosterone The principal male *sex hormone* (androgen), which stimulates and controls the development of *sperm* and of some *secondary sexual characteristics* (e.g. growth of facial hair and development of muscles). Testosterone is also produced in small amounts in females.

tetanus 1. Sustained contraction of *muscle* due to the fusion of small contractions (twitches) from many muscle fibres.
2. Lockjaw, an often fatal disease of humans, caused by the bacterium *Clostridium tetani*, that results in extreme muscle stiffness and rigidity requiring prolonged and painful treatment.

theory A general explanation of a wide variety of connected phenomena, based on and supported by extensive observations and experimental evidence. If any evidence contradicts the theory, the theory should be modified or rejected.

thermodynamic laws Four laws concerned with the relationship between usable energy, heat, and mechanical work within a closed system. The zeroth law states that if two bodies are at the same temperature as a third body, all three bodies are in thermal equilibrium. The first law (the law of conservation of energy) states that energy can be neither created nor destroyed, so that when one form of energy is transformed into another there is no energy loss or gain. The second law (the law of entropy) states that when one form of energy is transformed to another a proportion of energy is turned to heat energy. In biological systems, this means that energy transformations result in a loss of usable energy. The third law states that at a temperature of absolute zero (0 K or -273.15 °C) there will be no change in entropy in a perfect crystalline solid.

thermoregulation Control of body temperature within a relatively narrow range by either behavioural means (see *ectotherm*) or metabolic means (see *endotherm*).

Thermoregulation is a form of *homeostasis* involving negative *feedback*.

thiamine See *vitamin B1*.

thorax 1. (in tetrapods) The chest cavity, containing the heart and lungs. In mammals, it is separated from the *abdomen* by the *diaphragm*.
2. (in *arthropods*) The part of the body behind the head and in front of the abdomen. The thorax of insects is divided into three segments, each of which has a pair of jointed legs.

thrombin See *blood clotting*.

thrombosis The formation of a blood clot, especially one inside a blood vessel that remains at its point of formation.

thylakoid A flattened, fluid-filled sac within a *chloroplast*. Each sac is lined with a cell membrane containing *chlorophyll* used in the *light-dependent stage* of *photosynthesis*. Stacks of thylakoids are called *grana*.

thymine See *pyrimidine*.

thyroid gland An *endocrine gland* in tetrapods, situated in the neck, that secretes *thyroxine* (or *thyroxin*), an iodine-containing *hormone* which increases the *basal metabolic rate* of most tissues. Deficiency of thyroxine during development causes *hypothyroidism* (reduced mental and physical development); excessive secretion of thyroxine causes *hyperthyroidism* (hyper-activity, swelling of the neck, and protrusion of the eyes, a condition also known as exophthalmic goitre).

thyroxin See *thyroid gland*.

thyroxine See *thyroid gland*.

tibia The larger of the two long bones in each hind limb of tetrapods. In humans, it is called the shin bone and articulates with the femur at the knee joint and with the foot at the ankle bone (tarsus).

tidal volume (in mammals) The volume of air inhaled into the lungs or exhaled out of the lungs during each breath. The typical resting tidal volume for humans is about 500 millilitres, but it increases dramatically during exercise.

tissue A group of similar cells and the intercellular substances associated with them that are physically linked together to perform one or more particular functions in an organism. Some tissues are all of one cell type (e.g. *parenchyma* in plants and ciliated *epithelium* in mammals), but others contain a mixture of different cells (e.g. *xylem* and *phloem* in plants).

tissue fluid The fluid bathing the cells and forming the internal environment in a multicellular organism. In vertebrates, tissue fluid is formed by *ultrafiltration* of blood plasma at the arterial end of a *capillary*. Consequently, its composition is very similar to that of blood plasma, but it lacks most proteins (including *fibrinogen*) and *red blood cells*.

T-lymphocyte or **T cell** A small *white blood cell* involved in immune reactions. Each type of T-lymphocyte has a special cell-surface receptor which enables it to recognize a specific *antigen*.

tongue A muscular organ in vertebrates, usually attached to the floor of the *buccal cavity*, used to manipulate and taste food before swallowing.

tonoplast The *partially permeable membrane* surrounding a plant *vacuole*, separating the cell sap from the *cytoplasm*; it controls the exchange of material between the *cytoplasm* and vacuole.

tooth One of the hard structures within the *buccal cavity* (mainly on the jaws) of vertebrates, that are used for acquiring, biting, tearing and crushing food before it is swallowed. Some animals have teeth of one kind only (*homodont dentition*), but mammals have different teeth (see *incisor*, *canine*, *carnassial*, *premolar* and *molar*) to carry out different functions (*heterodont dentition*). The exposed part of each tooth is covered with enamel. The bulk of the tooth is made of *dentine* (a substance similar to bone but with a higher mineral content and no cells); the centre is made of soft tissue (*pulp*) which contains *blood vessels* and *nerve fibres*; and the base (*root*), set in the jaw, is covered in cement and embedded in the gums. See also *dental formula*.

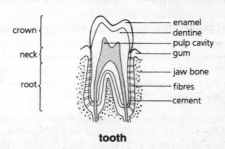

tooth

top carnivore An animal that preys on other animals but which is not itself preyed upon.

toxin A poisonous *protein* produced by bacteria, plants or animals; toxins act as *antigens*.

trace element An element required by living organisms in minute quantities to ensure normal growth and development. Trace elements essential in the human diet include iodine and selenium. Trace elements essential for plants, without which they die, include boron, copper, manganese, molybdenum and zinc.

trachea 1. (in tetrapods) The windpipe leading from the *pharynx* and carrying air to the lungs.
2. (in insects) A tube through which air enters the body. See also *tracheal system*.

tracheal system A branching system of air tubes in insects, which allows gaseous oxygen to diffuse from the external environment directly to respiring tissues. Small openings on the body surface (spiracles) allow air to enter cavities in the system. Extending from these are tubes (trachea) kept open by spiral or annular patterns of *chitin*; each trachea branches into numerous small tubules (tracheoles) which penetrate internal tissue. Tracheoles are very thin and not lined with chitin. Their walls provide the surface through which *gas exchange* takes place.

tracheid See *xylem*.

tracheole See *tracheal system*.

tracheophyte A plant that has a conspicuous *sporophyte* generation, elaborate vascular tissues (*xylem* and *phloem*), and complex leaves with waxy cuticles. In some classifications the tracheophytes are ranked as a phylum (or division). It is generally recommended that the term tracheophyte should be used as a collective noun to describe the phyla (divisions) Pteridophyta (clubmosses, ferns, etc.) and Spermatophyta (seed plants). *Filicinophyta* (ferns), *Coniferophyta* (conifers) and *Angiospermatophyta* (flowering plants) are tracheophytes.

trait See *character*.

transamination The process by which amino groups are transferred from an *amino acid* to a keto acid to form a new amino acid and keto acid (e.g. an acid such as *pyruvate* that contains a keto group, C=O). Transamination is important in the breakdown of excess amino acids and in the formation of new amino acids. Non-essential amino acids can all be formed by transamination but the essential amino acids can be obtained only from the diet.

transcription The synthesis of a strand of *RNA* from a single strand of *DNA*. The double helix of a DNA molecule unwinds to expose a single strand; this acts as template for the synthesis of messenger RNA, transfer RNA, or ribosomal RNA. Prior to protein synthesis, the DNA *genetic code* for a polypeptide is transcribed into mRNA. The sequence of bases on DNA determines the sequence of bases in RNA by complementary base pairing. A DNA base sequence of adenine, cytosine, guanine and thymine is transcribed into an RNA base sequence of uracil, guanine, cytosine and adenine. In addition to DNA, transcription requires the appropriate *enzymes* (RNA polymerases) and energy in the form of *ATP*.

transduction 1. The transfer of DNA from one bacterium to another using a *virus* as a *vector*.
2. The transformation by a *receptor cell* of an environmental stimulus into a *nerve impulse*.

transect A line or strip used in ecology for sampling organisms or environmental factors. A line transect consists of a single line; all organisms touching the line are recorded. A belt transect consists of a broad strip inside which the organisms are recorded. The transect may be continuous (the whole line or belt is sampled) or interrupted (samples are made at random or selected, usually at regularly spaced points). Transects are particularly useful for studying the transition from one community to another in *ecosystems* such as sand dunes and rocky shores.

transfer RNA See *RNA*.

translation The process by which the genetic information encoded in mRNA directs the synthesis of proteins. *Ribosomes* use the mRNA as a template for the synthesis of individual polypeptide chains. A ribosome moves along an mRNA molecule 'reading' each *codon* and matching it with a complementary anticodon in tRNA that carries a specific *amino acid*; thus the sequence of codons in mRNA determines the sequence of amino acids in the polypeptide chain.

translocation The transport of organic solutes (e.g. *sucrose*, *amino acids* and *plant growth substances*) in the *phloem* from one part of a tracheophyte plant to another. Translocation is an active process requiring energy (*ATP*) and occurring only in living plants. No single theory offers a fully satisfactory explanation of the mechanism of translocation. It may involve *mass flow*, *cytoplasmic streaming*, a special type of *osmosis* (electro-osmosis), or a combination of these. In electro-osmosis, a change in the electrical charge (potential difference) across a sieve plate is brought about by *active transport* of potassium ions. This is thought to result in the movement of charged water molecules and their solutes from one side of the sieve plate to the other.

transmission Transport from one place to another; for example, of a *nerve impulse* along a *nerve fibre* and from one *neurone* to the next (see *synapse*), and of a parasite from one host to another.

transmission electron microscope See *electron microscope*.

transpiration The diffusion of water vapour from aerial parts of a plant into the surrounding atmosphere. Transpiration occurs mainly through stomata (see *stoma*), but some water is also lost through other parts of the plant, especially the *lenticels*. When water molecules leave the plant by transpiration, they exert a strong upward pull on water molecules remaining in *xylem* vessels because of *cohesion*. This is thought partly to explain the continuous flow of water through a plant from roots to leaves (known as the transpiration stream). Another contributory factor is *adhesion* between water molecules and the walls of xylem vessels which prevents the molecules from slipping downwards. Transpiration tends to be highest on hot, dry, windy and sunny days when there is a plentiful supply of water to the plant. Most losses of water by transpiration are probably an inevitable result of stomata being open for carbon dioxide uptake during *photosynthesis*. Transpiration may also assist cooling and concentrating nutrients around the roots. Less than 1% of the *transpiration stream* is used in photosynthesis and growth.

transpiration stream See *transpiration*.

transverse (of a section) Cut at right angles to the longitudinal axis of an organism or organ.

triacylglycerol A member of a group of neutral fats and *oils* formed by the addition of three *fatty acid* groups to *glycerol*.

tricarboxylic acid cycle An alternative name for the Krebs cycle. See *aerobic respiration*.

trichromatic theory See *vision*.

tricuspid valve See *heart*.

triglyceride The former name for *triacylglycerol*.

triose A *monosaccharide* containing three carbon atoms in each molecule.

triose phosphate A three-carbon sugar to which a phosphate group has been added (i.e. it has been phosphorylated). GALP is a triose phosphate.

triplet See *codon*.

triploblastic (of an organism) Having three body layers in the *embryo*, from which all cells are derived. The layers are the *ectoderm*, *endoderm* and *mesoderm*.

trivial name See *binomial nomenclature*.

tRNA See *RNA*.

trophic level A feeding level in an *ecosystem*; a stage in a *pyramid of energy*, biomass or numbers, consisting of organisms of similar nutritional type through which food and energy flow. Producers form the first trophic level, primary consumers the second, and so on. Decomposers belong to no particular trophic level, because they consume dead organic material derived from all levels.

tropism A movement (almost always a growth movement) by part of a plant in response to an external stimulus and in a direction related to the stimulus. *Chemotropism* is a response to a chemical stimulus (e.g. *pollen tubes* exhibit positive chemotropism when they grow towards chemicals produced at the *micropyle* of an ovule). *Geotropism* is a response to the

Earth's gravity; the main root of a flowering plant is usually positively geotropic (growing downwards) and the main stem negatively geotropic (grows upwards). *Hydrotropism* is a response to water; roots and pollen tubes are positively hydrotropic (grow towards water). *Phototropism* is a response to light; shoots and *coleoptiles* are positively phototropic (grow towards light) and some roots (e.g. the adventitious roots of climbers such as ivy) are negatively phototropic (grow away from the direction of light). Tropisms are thought to be controlled by *plant growth substances* (e.g. in a shoot, auxins are thought to move laterally away from the light, resulting in a higher concentration of auxin on the shaded side, thus causing elongation of cells and curvature of the shoot towards the light).

trypsinogen An inactive precursor of trypsin, secreted by the *pancreas*.

t-test A statistical test of the significance of the difference between the *means* of two samples (sample size less than 30) with measurements made at the interval level (i.e. measurements such as height in centimetres in which the size of differences between measurements can be assessed).

tuber A swollen, underground plant stem or root which acts as a perennating organ (see *perennation*) and may also provide a means of *vegetative reproduction*. In potatoes, tubers form at the end of the stem, each tuber incorporating several *nodes* and internodes, with buds in the axils of tiny leaves (the 'eyes') that give rise to a new plant. Dahlia tubers develop from swollen adventitious roots.

tubulin See *microtubule*.

Tullgren funnel A device used to extract small invertebrates (especially arthropods) from samples of soil or leaf litter. Dry material is placed in a funnel with a perforated opening (e.g. gauze) at the bottom to hold the soil in place but also to allow organisms to pass through. A 5-watt lamp is shone above the soil. Soil organisms move away from the heat and light, through the perforations, and drop into a collecting vessel beneath the funnel which may contain preservative. Usually, most specimens are collected in two hours, but complete extraction may take several days.

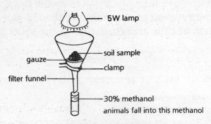

Tullgren funnel

turgid (usually of a plant cell) Incapable of taking up more water because of resistance by the cellulose *cell wall*. Fully turgid cells are rigid and make a significant contribution to the mechanical support of plant tissues. See also *pressure potential*; compare *flaccid* and *wilting*.

turgor pressure See *pressure potential*.

twins Two offspring born at the same time from the same mammalian mother. Non-identical twins (sometimes called *dizygotic* or fraternal twins) develop from different eggs. Identical twins (sometimes called *monozygotic* twins) develop from the same fertilized egg.

tympanic membrane or **eardrum** The membrane which transmits sound to the middle *ear*.

ulna A long bone in the forearm of vertebrates, which joins the *humerus* at the elbow joint. In humans, it is larger and lies further away from the thumb than the radius. In some vertebrates, the ulna is combined with the radius to form a single bone.

ultracentrifugation A technique for separating large molecules and cell *organelles* in a suspension, by spinning the suspension in a tube at various speeds up to 150 000 g (where g is the acceleration due to gravity and has a value of 9.8 m s^{-2}). Each speed causes the sedimentation of a particular molecule or organelle; the smaller the molecule or organelle, the higher the speed required for its sedimentation. See also *subcellular fractionation*.

ultrafiltration The process by which small molecules and ions in the blood are separated from larger molecules as they pass through the basement membrane of capillaries to produce either the glomerular filtrate (see *glomerulus*) or the *tissue fluid*.

ultrastructure The detailed structure of a cell as seen with an *electron microscope*.

umbilical cord A cord consisting of *blood vessels* and *connective tissue* which connects the *embryo* to the *placenta* in most pregnant mammals, carrying nutrients and oxygen to the embryo, and waste products away from it.

undernourishment A form of *malnutrition* in which there is either an inadequate amount of food, or a deficiency of one or more nutrients.

undulipod A single-celled *eukaryote* that has *cilia* or *flagella*; the group includes *Paramecium* and *Euglena*.

unicell See *unicellular organisms*.

unicellular organism or **unicell** An organism consisting of a single cell. Unicellular organisms include bacteria, protozoa, and certain algae and fungi (e.g. yeast). They are often placed in a separate kingdom, the *Protoctista*.

units of measurement See appendix 4 (SI units).

unsaturated fatty acid A *fatty acid* which has at least one double covalent bond within the hydrocarbon chain. This imposes a kink in the chain, making molecules of *triacylglycerols* made from unsaturated fatty acids less easy to pack together and more fluid. Unsaturated fatty acids tend to form *oils* (olive oil, for example, contains high concentrations of oleic acid).

uracil See *pyrimidine*.

urea An organic molecule which is the main nitrogen-containing excretory product of mammals. It is formed in the *liver* from ammonia and carbon dioxide in the *ornithine cycle* and is eliminated from the body in the urine and, to a lesser extent, in sweat. Urea (formula $NH_2.CO.NH_2$) was the first organic molecule to be synthesized in the laboratory (in 1832) and later industrially.

urea cycle See *ornithine cycle*.

ureter One of a pair of tubes which carry urine from each kidney to the bladder.

urethra The tube that carries urine from the bladder to the exterior. In male mammals, it also conveys *semen*.

urinary system The structures associated with the *kidney* and involved in the elimination of urine from the body. In mammals, these include the kidneys, *ureters*, *bladder* and *urethra*. The urinary system is often closely linked with the reproductive system and the two are some times referred to as the urinogenital system.

urine A watery solution of *urea* and mineral salts produced by the *kidney* in mammals. Urine is stored in the *bladder* and discharged into the external environment through the *urethra*.

urinogenital system See *urinary system*.

uterus or **womb** In female mammals, the muscular cavity in which an *embryo* can develop. It receives egg cells from the *oviduct* and is connected posteriorly to the *vagina*. The size of the uterus and the thickness of its walls vary cyclically (see *menstrual cycle*). During pregnancy it becomes much more muscular so it is able to expel the young through the vagina at birth.

vaccine See *immunization*.

vacuole A fluid-filled, membrane-bound sac within a cell. A plant cell vacuole contains cell sap.

vagina The tube leading from the *uterus* to the external environment in female mammals. Its walls are muscular and lined with mucus glands. The vagina receives the *penis* during mating and *sperm* are introduced through it into the uterus, where *fertilization* may take place. During birth, fully developed foetuses leave the mother through the vagina.

valve A membranous structure that restricts the flow of fluid to one direction only through an aperture or tube. For example, in the *circulatory system* of vertebrates, valves allow blood to flow around the body in only one direction.

vapour pressure The pressure exerted by a vapour, either by itself or in a mixture of gases.

variable A changeable aspect of a situation which can be manipulated or measured. In an investigation, an independent variable is one whose values are controlled in order to observe its effects on a dependent variable. Some investigations involve two dependent variables which act on and influence each other.

variance A measure of variation about an arithmetic *mean*, in statistics, calculated as the average squared deviation of all individual values from the mean value

$$s^2 = \Sigma(x - \bar{x})^2/n,$$

where s^2 is the sample variance, Σ is the sum, x is an individual measurement, $\bar{x}$ is the mean of individual measurements, and n is the number in the sample. The square root of the variance is called the *standard deviation*.

variation Differences in the *phenotypes* of individuals belonging to the same species that may be acted on by *natural selection*. The differences may be due to genetic factors, environmental factors, or a combination of both. *Mutations* are the ultimate source of inheritable, genetic variations. Characteristics that have a *continuous variation* (or *quantitative variation*) (such as height in humans) show slight differences that grade into each other. They are usually determined by a large number of genes or influenced greatly by environmental factors. Characteristics that have a *discontinuous variation* (also called *qualitative variation*) fall into discrete groups that do not overlap (for example, garden peas are either wrinkled or smooth). They are often controlled by one or a small number of genes.

vascular bundle A structure in *tracheophyte* plants, consisting of a long, continuous strand of vascular tissue (mainly *xylem* and *phloem* supported by *sclerenchyma* and *collenchyma*) extending from the roots to the leaves. In flowering plants, the vascular bundles are visible as veins on the surface of leaves.

vascular system A specialized system of tubes and vessels for the transport of fluid within an organism. In mammals, there are two main vascular systems: the blood vascular system consisting of the heart blood vessels (arteries, capillaries and veins), and the lymphatic system. In Echinodermata, a special system of water-filled tubes (the water vascular system) transmits forces through tube feet for locomotion. In *tracheophyte* plants, the vascular system consists of vascular tissue, mainly *xylem* and *phloem*, forming a continuous system of vessels that carries water, mineral salts and food nutrients, and also provides mechanical support.

vascular tissue See *vascular system*.

vector 1. A carrier (e.g. *plasmid* or *bacteriophage*) of DNA from one organism to another, or of a **parasite** from one host to another.
2. A quantity that can be described only by using two or more numbers (e.g. magnitude and direction), or a matrix with only one row or column of data (called components).

vegetative propagation See *vegetative reproduction*.

vegetative reproduction or **vegetative propagation** Asexual reproduction in which part of a plant that includes a shoot apex or bud detaches from the parent plant and develops into a new individual. Plant structures which can undergo vegetative reproduction include *bulbs*, *corms*, *rhizomes*, *runners* and *tubers*.

vein 1. A *blood vessel* with thin walls and a relatively large lumen that carries blood towards the heart, in animals. Valves in the vein ensure that blood flows in one direction only. Compare *artery*.
2. In plants, a *vascular bundle* in a leaf.

vena cava One of the major *veins* in the *circulatory system* of vertebrates, which conveys blood to the right atrium of the *heart*. In mammals, a superior vena cava (also called anterior vena cava) carries blood from the head, neck, and upper limbs and an inferior vena cava (also called posterior vena cava) carries blood from the rest of the body (except the lungs) and lower limbs.

ventilation or **breathing** Rhythmic movements by which a respiratory medium (air or water) is pumped into a respiratory organ (inspiration) and out of it again (expiration) so that *gas exchange* can take place. Air or water moves from a high pressure to a low pressure. Breathing or ventilatory movements of the body ensure that during inspiration the pressure in the respiratory organ is lower than that in the environment, and that the reverse happens during expiration. For example, in mammals, ventilation involves the combined action of the *ribs* and *diaphragm*; during inspiration the ribs move upwards and outwards and the diaphragm contracts downwards, increasing the volume of the thoracic cavity and decreasing the pressure around the lungs, sucking air inwards; movements in the opposite directions occur during expiration.

ventral 1. (in vertebrates) Pertaining to the surface of the body furthest from the notochord or backbone.
2. (in invertebrates) Pertaining to the surface of the body closest to the nerve cord.
3. (in organisms with no backbone or nerve cord) Pertaining to the surface of the body that is normally closest to the substrate.

ventral root One of the points at which *nerve fibres* of motor *neurones* leave the ventral surface of the spinal cord or brain.

ventricle See *heart*.

vertebra One of the bones or segments of *cartilage* that form the vertebral column (or backbone) in vertebrates, and from which the group acquires its name. Each vertebra consists of a solid body (the *centrum*) which extends upwards to form an arch (the *neural arch*) enclosing and protecting the spinal cord. A dorsal extension of the arch (the *neural spine*) acts as an important attachment point for locomotory muscles. Other extensions of bone or cartilage also act as muscle attachment points, or prevent too much movement between adjacent vertebrae. A cartilaginous disc (the *intervertebral disc*) containing a semi-fluid material acts as a cushioning pad between adjacent vertebrae.

vertebral column, backbone or **spinal column** In vertebrates, the series of vertebrae which form a longitudinal rod down the length of the body, from the cranium to the tail. It is the main supporting structure of the body. It is flexible, and muscles attached to it are important for locomotion. It also supports and protects the *spinal cord*.

Vertebrata A subphylum of the *Chordata*. See *vertebrate*.

vertebrate An animal with a backbone enclosing a *spinal cord*; a member of subphylum *Vertebrata*. The brain is enclosed within a skeletal case, the cranium.

vesicle A small *vacuole*.

vibrios A group of bacteria that have a single *flagellum*, giving them a comma-like shape. An example is *Vibrio cholerae*, which causes cholera.

villus (*pl.* **villi**) A finger-like extension (e.g. from the *small intestine* or *placenta*). The vast number of villi in the small intestine increase the surface area of the gut wall for absorption of nutrients. Blood vessels in the villus absorb water-soluble products of digestion (e.g. *glucose* and *amino acids*) and *lacteals* absorb fats. See also *duodenum*.

virgin birth See *parthenogenesis*.

virus A minute, infectious particle that can pass through a fine filter. Viruses range in size from 20 to 300 nm and can be seen only with the aid of an electron microscope. They consist of a length of genetic material (either DNA or RNA) which forms a core inside a protein coat. Some viruses have an additional *lipid* and glycoprotein coat. Viruses have no metabolism of their own and can reproduce only inside a host cell. They control the metabolism of host cells and use its *enzymes*, raw materials, and *organelles* to reproduce. New viruses are usually released when the host cell disintegrates. Viruses cause many diseases of plants and animals. Viruses that infect bacteria are called *bacteriophages*. See also *reverse transcriptase*.

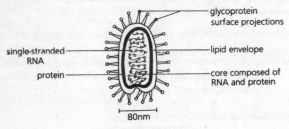

glycoprotein
surface projections

single-stranded RNA

lipid envelope

protein

core composed of RNA and protein

80nm

virus The HIV virion.

viscera The soft, internal organs of an animal, collectively.

vision The ability to detect and process reflected light in order to perceive an image of the object from which the light is reflected. Detection of light takes place in the *eye*, but perception, the process of forming an image and 'seeing' an object, takes place in the brain. Some vertebrates perceive a coloured image (colour vision) by means of cone cells in the retina; there are three types of cone cells containing pigments sensitive to red, blue, and green light and it is thought that colours are detected by the relative degree of stimulation of the three types of cone (the so-called *trichromatic theory* of colour vision). Perception of white would then be due to equal stimulation of all three types. Animals lacking colour vision probably form images of objects in shades of grey. The ability to perceive three-dimensional images, facilitating judgement of distances, is called *binocular vision* and occurs in vertebrates able to view an object using both eyes simultaneously; this causes the object to be viewed from slightly different positions, creating a stereoscopic effect. Humans have binocular vision, but it is poor in comparison to such predators as owls and cats.

vital capacity The maximum volume of air that can be expired after a maximum forced inspiration. It is a measure of the maximum volume of air which can be exchanged in the lungs during one breath. In humans, values vary from about 3 to 6 litres.

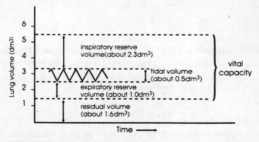

vital capacity The vital capacity of a man at rest.

311

vitamin An organic compound that is required in minute quantities by heterotrophs for their normal health and development, but which they cannot synthesize themselves. Vitamins are non-protein substances which are required for energy transformations and the regulation of *metabolism*; lack of vitamins results in deficiency diseases. Until early this century, it was believed that a diet with adequate amounts of protein, carbohydrate, and fat was sufficient to maintain health, but in 1912, F.G. Hopkins published a paper questioning this. His experiments on nutrition in rats suggested there were other factors essential to growth and health. He called these 'accessory growth factors'. His paper stimulated research workers to isolate these factors. So far, 13 compounds have been classed as vitamins. Many vitamins, such as *vitamin B1* (thiamine) may function as coenzymes (see *cofactor*).

vitamin A or **retinol** A fat-soluble *vitamin* which helps the *mucous membranes* of the eye and respiratory tract to function properly. Vitamin A is also used in the formation of visual pigments in the eye. It can be manufactured in the body from beta-carotene which is found in a variety of foods, particularly green vegetables, carrots, and fish-liver oils. Vitamin A deficiency increases the risk of infections of the respiratory, digestive and urinogenital tracts, and causes a number of eye disorders, including night blindness, and even permanent blindness (xerophthal mia).

vitamin B complex A group of water-soluble *vitamins* that play an essential role as coenzymes (see *cofactor*) in *respiration*. The group includes biotin, folic acid, *niacin*, *pantothenic acid* and vitamins *B1*, *B2*, *B6* and *B12*.

vitamin B1 or **thiamine** A water-soluble *vitamin* first obtained from rice polishings; it is also found in lean meat, liver, eggs, whole grains, and milk. It plays a very important role in releasing energy from foods rich in *carbohydrates*. Gross deficiency causes the potentially fatal disease *beriberi*. Mild deficiencies cause fatigue, loss of appetite, muscle weakness, and digestive disturbances. Vitamin B1 is rapidly destroyed by heat and is stored in only very limited amounts in the body.

vitamin B2 or **riboflavin** A water-soluble *vitamin* quickly decomposed by heat. Riboflavin is obtained from a wide variety of foods including whole grains, yeast, liver, eggs and milk. It forms coenzymes (see *cofactor*), such as flavoprotein, which play a central role in releasing energy from food. Deficiency causes ariboflavinosis, characterized by cracked skin and ocular problems including blurred vision.

vitamin B3 See *niacin*.

vitamin B5 See *pantothenic acid*.

vitamin B6 A complex, water-soluble *vitamin* obtained from meat, poultry and eggs. It plays a role as a coenzyme (see *cofactor*) in *amino acid* and *carbohydrate* metabolism. It is also involved in *antibody* formation and *haemoglobin* synthesis. Although rare, deficiency causes nervous irritability, anaemia and convulsions.

vitamin B12 or **cobalamin** A water-soluble *vitamin*, containing cobalt, required for the formation of red blood cells. Deficiency causes pernicious anaemia. It can be obtained from liver and fish, but most vitamin B12 is synthesized in the gut by symbiotic bacteria.

vitamin C or **ascorbic acid** A water-soluble *vitamin* essential for the formation of *collagen* (a major component of skin, muscles, and bone) and the healthy functioning of tissues containing collagen. It is required for the repair of joint tissues which are often damaged during high levels of physical activity. Vitamin C acts as a stimulant of the body defence mechanisms and protects vitamin A, vitamin E, and dietary fats from oxidation. Vitamin C also plays an important role in the absorption of iron from non-animal sources. Mild deficiencies can cause fleeting joint pains, poor tooth and bone growth, poor wound healing, and an increased susceptibility to infection. Extreme deficiency causes scurvy, characterized by impaired wound healing, bleeding gums, and loose teeth. Vitamin C is obtained from citrus fruits, green vegetables and tomatoes.

vitamin D A fat-soluble *vitamin* containing a number of chemicals that enhance the absorption of calcium and phosphorus from the intestine and, with parathyroid hormone, mobilizes their deposition in bones. Vitamin D is relatively stable when exposed to heat and light. It is stored in the liver and, to a lesser extent, in the fatty tissue in the skin. Vitamin D occurs in two forms: vitamin D2 and vitamin D3. Vitamin D2 (or ergo-calciferol or calciferol) is obtained in the diet from foods such as oily fish, eggs, and margarine. Vitamin D3 (or cholecalciferol) is the main form of the vitamin and is produced in the skin by the action of ultraviolet light on 7-dehydrocholesterol. Vitamin D is also obtained from fish-liver oils, milk, and egg yolk. Deficiency causes a loss of muscle tone, restlessness and irritability; it also causes rickets in children, and demineralization and softening of the bones (osteomalacia) in adults.

vitamin E A group of related compounds called tocopherols, believed to maintain the integrity of *cell membranes* and to act as an anti blood-clotting agent. Vitamin E is widely available in the diet. The richest sources include wheatgerm oil, sunflower oil and roasted peanuts. Deficiencies are rare, but when they do occur they may lead to destruction of red blood cells and anaemia. Deficiency impairs reproductive ability in rats and causes muscle wasting in pigs.

vitamin K A fat-soluble *vitamin*, dietary sources of which include green leafy vegetables, some vegetable oils, and liver. About half of the human vitamin K requirement is derived from intestinal bacteria synthesis. Vitamin K is needed to manufacture prothrombin, which is essential for the normal clotting of blood. It is also thought to be involved in *oxidative phosphorylation*. Deficiency is rare but may result from taking antibiotics which interfere with the activity of gut bacteria. Deficiency symptoms include easy bruising and prolonged clotting time leading to excessive bleeding and haemorrhage.

vitreous humour See *eye*.

viviparous See *birth*.

vocal cords See *larynx*.

voluntary muscle See *skeletal muscle*.

vulva The external opening of the *vagina*. It forms the external genitalia of female mammals, comprising two fleshy folds of tissue (the *labia*).

wall pressure See *pressure potential*.

waste product A by-product of *metabolism* that has no beneficial function. Harmful waste products (e.g. ammonia and *urea*) must be eliminated (excreted) from the body. See *metabolic waste*.

water A colourless, odourless liquid that is the most abundant molecule in living organisms: 60—90% of living matter is water; 75% of the Earth's surface is covered with water which provides an important external environment for organisms. Water has a number of properties that make it an excellent internal and external environment. It is called the 'universal solvent', because it can dissolve more substances than any other liquid. *Sugars*, *salts*, gases, *amino acids* and small *nucleic acids* all dissolve relatively easily in water. This enables it to act as a good transport medium, and many chemicals react more efficiently when in solution. Water has a high *specific heat capacity* (i.e. it must absorb a relatively large amount of heat before its temperature rises). It has a high *latent heat of vaporization*, so a large amount of heat must be absorbed to turn liquid water into water vapour and water is therefore liquid over a much wider range of temperatures than other similar molecules. Its high latent heat of vaporization also enables water to be used as a coolant.

Water has peculiar freezing properties. It is most dense at 4 °C and as it cools below this temperature its volume increases and density decreases (which is why ice tends to float rather than sink). The coldest water lies at the surface, acting as an insulating layer for the water below and allowing organisms that live at the bottom of aquatic habitats to escape freezing. Water has the highest *surface tension* of any liquid, a property responsible for its forming bubbles and drops, and for *capillarity*, and its very high surface tension makes the surface of water behave like an elastic skin supporting the weight of organisms such as pond skaters. Most of these properties are due to the ability of water molecules to link together by *hydrogen bonding*. Water has other properties important to life and plays a central role in some physiological processes. It is the source of hydrogen for *photosynthesis*, is used to digest foods by *hydrolysis*, and because it is relatively incompressible it is used in fluid-filled cavities as a *hydrostatic skeleton*. Its transparency allows light to penetrate aquatic habitats so that photosynthesis can take place.

water budget The balance of water gained and lost in a biological system. The water budget of a terrestrial mammal depends on a number of factors: gains include drinking water, water in food, and water produced by *respiration* of fats and carbohydrates; losses include sweat, urine, faeces, and water vapour in air exhaled from the respiratory tract. Small variations in any one of these factors may be critical for desert animals.

water cycle or **hydrological cycle** The flow of water as liquid and vapour through the *biosphere*. The cycle is driven by solar and gravitational energy. On a global scale, the amounts of water used in *photosynthesis* and produced in *respiration* are so small they can usually be ignored. The major contribution made by non-human organisms to the water cycle is by *evapotranspiration*. Humans accelerate the cycle by heating water, for domestic and industrial purposes, thus increasing the rate of evaporation, and by burning fuels containing carbon and hydrogen (called hydrocarbons) which release water vapour as a combustion product. On average, each molecule of water spends 10—15 days between entering the atmosphere as vapour and returning as precipitation.

water potential (ψ) A measure of the pressure exerted by water molecules on a membrane when a solution is separated from pure water and the membrane is permeable to water only. The water potential of a system (e.g. a cell) is the difference between the chemical potential of the water in the system and the chemical potential of pure water at the same pressure. The water potential of pure water at atmospheric pressure is taken to be zero. Solutes reduce the water potential of a solution; all solutions therefore have a water potential of less than zero. Water tends to move from a region of higher (less negative) water potential to a region of lower (more negative) water potential. Water potential is usually measured in kilopascals (kPa). See also *osmosis*.

water potential gradient The difference in *water potential* between two regions. It is a form of *concentration gradient*. Water always moves down a water potential gradient.

weather The atmospheric conditions (especially humidity, precipitation, air temperature, wind speed and direction, and atmospheric pressure) experienced in a particular place from day to day. Compare *climate*.

weathering The breakdown of rocks and minerals near the ground surface by the action of wind, water, ice, chemical reaction, or biological activity. Weathered rock particles form the main bulk of *soil*. The transport of weathered material from its site of formation is called *erosion*.

weed A plant growing in what humans consider the wrong place that has some (usually economic) detrimental effect; a plant pest. Many types of weed compete for resources with plants considered beneficial, but the same plant may, under different circumstances, be prized and itself cultivated.

white blood cell or **leucocyte** A blood cell that contains no *haemoglobin* or other respiratory pigment. White blood cells are important in defending the body against disease. Some produce *antibodies*, some engulf bacteria and foreign bodies, and some produce antitoxins to neutralize poisons. White blood cells can squeeze through *capillaries* and enter any tissue, including bone.

white matter Part of the *central nervous system*, lying outside *grey matter*, that contains mainly *nerve fibres* and few cell bodies. Myelin encapsulating the fibres is responsible for its white appearance.

wild type An individual carrying normal (unaltered) alleles of one or more *genes*, as contrasted with one with *mutant* genes that may be able to survive only under laboratory conditions. The term is often used by those studying the genetics of *Drosophila*. Use of the term 'wild type' and its symbol (+) can be confusing and it is generally recommended that students use alternative descriptions.

wilting Drooping and loss of rigidity in plants. It usually occurs when *transpiration* losses exceed water absorption. Excessive wilting may be harmful and irreversible.

wind The movement of air parallel to the ground surface.

windpipe See *trachea*.

wing An organ of flight. A bat wing and bird wing are both modified *pentadactyl limbs*. The wing of an insect is a membranous outgrowth of the *thorax*.

womb See *uterus*.

wood 1. *Xylem*, formed as a result of *secondary thickening*, which has matured and secreted thick walls. Wood helps to strengthen trees and shrubs.
2. An area of land in which trees are growing.

X chromosome A sex *chromosome* that occurs in pairs in the body cells of the homogametic sex (female in humans and *Drosophila*). The heterogametic sex carries one X chromosome and one *Y chromosome*. The X chromosome carries numerous genes in addition to those which help determine sex. See also *sex determination* and *sex linkage*.

xeromorphic characteristics See *xerophyte*.

xerophyte A plant that is adapted to live under conditions of water scarcity. Xerophytes have a number of structural features (called xeromorphic characteristics) that adapt them to their environment. Some cacti maximize water retention by means of thick, succulent stems and thick *cuticles*, and minimize water loss through *transpiration* by means of leaves reduced to spines, compensating for the loss of leaf area with green stems in which *photosynthesis* takes place. Marram grass, which inhabits the dry environment of sand dunes, has a thick, waxy cuticle and leaves which can curl inwards when dry; the *stomata* are sunk into pits on the inner surface of the leaf and become hidden when the leaf curls. Compare *halophyte*, *hydrophyte* and *mesophyte*.

xylem Woody plant tissue, consisting of a number of different cell types (all known as tracheids), including *sclerenchyma* fibres that provide mechanical support and xylem vessels, whose main function is to transport water during *transpiration*. Xylem vessels are formed from the fusion of the end walls of several cells to form long, empty tubes. The addition of *lignin* to cellulose cell walls makes the vessels impermeable, thicker and stronger, but the cells die.

Y chromosome A sex chromosome that occurs singly in the body cells of the heterogametic sex (male in humans and *Drosophila*). The Y chromosome carries few (if any) *genes*. See also *sex determination* and *sex linkage*.

yeast One of a group of unicellular *fungi* that are very important in a number of industrial processes, including brewing and baking. Most yeasts are facultative anaerobes that reproduce by *binary fission*. Under anaerobic conditions, they produce ethanol and carbon dioxide.

yellow body The *corpus luteum* formed from the *Graafian follicle* after *ovulation*.

yolk The food store in the majority of animal eggs. It consists mainly of protein and fat which nourishes the developing *embryo*.

Z-membrane A protein band which defines the boundary between one *sarcomere* and the next in a muscle fibre. The z is an abbreviation for *zwischen*, which, in German, means between.

zonation The spatial distibution of organisms into distinct zones with different environmental conditions. The term may be used of the broad distribution zones of plants, determined primarily by climatic factors, or the more local zonation within an *ecosystem*. Ecosystem zonation is particularly clear on rocky shores, where different species occupy distinct horizontal strips, roughly parallel to the water's edge, from the low tide mark to the high tide mark. Zonation superficially resembles succession, but in zonation the distribution of species varies in space (spatially) and in succession it varies in time (temporally).

zooplankton See *plankton*.

zwitterion An *ion* carrying both a positive and a negative charge. The charges cancel each other out and zwitterions are, therefore, electrically neutral. Zwitterions form from molecules, such as *amino acids*, that have both basic and acid groups. The electrical charge on an amino acid varies with *pH*. The pH at which the zwitterion is formed is called the *isoelectric point*.

zygote A *diploid* cell formed by the fusion of two *haploid* gametes at fertilization. One of the *gametes* (the male) is usually motile and smaller than the other, immobile gamete (the female).

Appendix 1

Graphs

Bar chart A diagram showing the relationship between two variables, one of which is not numerical. For example, a bar chart showing the percentage of vitamin C in different fruits consists of narrow rectangular blocks one representing each type of fruit. The height of each block (or bar) is proportional to the vitamin C content. The blocks may be arranged in any order, but they do not touch.

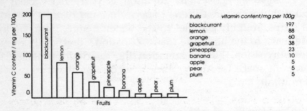

bar chart Table and bar chart showing the vitamin C content, in mg per 100 g, of nine fruits.

Column graph A graph showing the frequency distributions of discontinuous or discrete data. For example, a column graph can be used to show the frequency of occurrence of nests with different numbers of eggs. Each number of eggs is represented by a rectangular block, the height of which is proportional to the number of nests which have that number of eggs. The blocks for each number of eggs per nest do not touch each other; they are labelled centrally and drawn in order of increasing or decreasing magnitude.

number of eggs per nest	number of nests
1	0
2	0
3	1
4	3
5	16
6	14
7	4

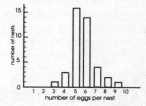

column graph Table and column graph showing the frequency of nests containing different numbers of eggs.

Histogram A graph showing the frequency (*y*-axis) of a continuous variable (*x*-axis). The data for the continuous variable are grouped into classes of equal width. Rectangular blocks, the area of which represents the frequency, are drawn in order of increasing or decreasing magnitude. The edges of the blocks are labelled, giving the common boundary points between adjacent classes rather than the classes themselves. Thus a block may be labelled '5' at the left and '6' at the right. Frequencies for '5' are included in this block, but frequencies for '6' are included in the next.

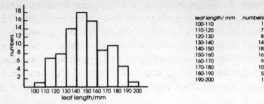

leaf length/ mm	numbers
100-110	1
110-120	7
120-130	8
130-140	14
140-150	18
150-160	16
160-170	9
170-180	10
180-190	5
190-200	1

histogram Table and histogram showing the frequency of different lengths of leaf.

Line graph A diagrammatic representation of the relationship, usually between two variables, in which only pure numbers are plotted. The independent variable (the variable whose value is chosen by the experimenter) is plotted on the x (horizontal) axis, and the dependent variable (e.g. readings taken on an instrument) is plotted on the y (vertical) axis.

The graph should have an informative title and the axes should be labelled with the appropriate quantity or unit of the variable. Points on the graph should be clear; dots are not generally regarded as suitable unless encircled. The line on the graph is called the curve, whether or not it is straight.

A smooth curve or single straight line should be drawn only if there is good reason to suppose that all the values between the plotted points fall on that curve. Straight lines drawn between successive points make no such assumption.

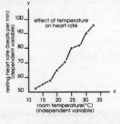

line graph

Pie chart A diagram used to compare various quantities, usually percentages or proportions of an identifiable whole.

The chart consists of a circle ('pie') marked into wedges corresponding in size with the various quantities to be compared. Different samples are compared using different pie charts, the size of which is proportional to the size of the sample, and the sequence of categories remains the same irrespective of their size.

If only one pie chart is drawn, the categories are usually drawn in rank order, beginning at '12 o'clock' and proceeding clockwise. Pie charts become difficult to interpret if there are more than about six sectors.

arthropod	sycamore	yew
barklice	31	15
beetles	8	18
bugs	8	2
flies	4	4
spiders	27	15
wasps	6	2

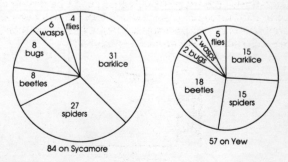

pie chart Table and pie chart showing the numbers of arthropods on two species of tree.

Appendix 2: geological time scale

Eon	Era	Sub-era	Period	Epoch	Began (millions of years ago)	
Phanerozoic	Cenozoic	Quaternary	Pleistogene	Holocene	0.01	Present time
				Pleistocene	2.0	Ice ages. Modern humans appear
			Neogene	Pliocene	5.1	First human ancestors
				Miocene	24.6	Forests in Europe and America. Modern mammals and birds
		Tertiary		Oligocene	38	Dogs, weasels, cats and rodents expanding
			Palaeogene	Eocene	55	Whales and sea cows return to the sea
				Palaeocene	65	Flowering plants become dominant on land. Sharks abundant
				Cretaceous	144	Flowering plants appear. Dinosaurs die out
	Mesozoic			Jurassic	213	Vast forests of conifers and tree ferns. Flies and butterflies appear. Dinosaurs reach great size
				Triassic	248	Mammals evolve. Deserts widespread

Eon	Era	Sub-era	Period	Epoch	Began (millions of years ago)	
				Permian	286	Reptiles dominant
		Upper Palaeozoic		Carboniferous	360	Amphibians abundant. First reptiles. Coal forests
	Palaeozoic			Devonian	408	Amphibians evolve. Fish evolve rapidly
				Silurian	438	First vascular plants and arthropods. Sea scorpions 2m long. Jawless fish abundant
		Lower Palaeozoic		Ordovician	505	Corals abundant. Reef-building algae. First jawless fish. First land plants
				Cambrian	590	Most invertebrate phyla present. First vertebrates. Marine algae abundant
	Sinian				800	First animals
	Riphean				1650	First eukaryotes
Proterozoic					2500	First photosynthesizing prokaryotes
Archaean					4000	First living organisms
Priscoan					4600	Earth probably lifeless

Note: Taken together, the Priscoan, Archaean and Proterozoic Eons are often called the Precambrian.

Appendix 3: taxonomy

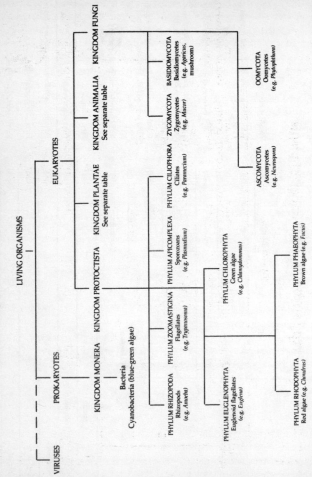

LIVING ORGANISMS

PROKARYOTES | EUKARYOTES

VIRUSES

KINGDOM MONERA
Bacteria
Cyanobacteria (blue-green algae)

KINGDOM PROTOCTISTA

KINGDOM PLANTAE
See separate table

KINGDOM ANIMALIA
See separate table

KINGDOM FUNGI

PHYLUM RHIZOPODA
Rhizopods
(e.g. Amoeba)

PHYLUM ZOOMASTIGINA
Flagellates
(e.g. Trypanosoma)

PHYLUM APICOMPLEXA
Sporozoans
(e.g. Plasmodium)

PHYLUM CILIOPHORA
Ciliates
(e.g. Paramecium)

PHYLUM EUGLENOPHYTA
Euglenoid flagellates
(e.g. Euglena)

PHYLUM CHLOROPHYTA
Green algae
(e.g. Chlamydomonas)

PHYLUM RHODOPHYTA
Red algae (e.g. Chondrus)

PHYLUM PHAEOPHYTA
Brown algae (e.g. Fucus)

ZYGOMYCOTA
Zygomycetes
(e.g. Mucor)

BASIDIOMYCOTA
Basidiomycetes
(e.g. Agaricus,
mushroom)

ASCOMYCOTA
Ascomycetes
(e.g. Neurospora)

OOMYCOTA
Oomycetes
(e.g. Phytophthora)

Appendix 3: classification of plants

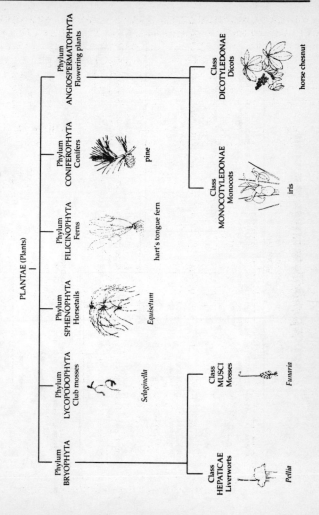

PLANTAE (Plants)

Phylum BRYOPHYTA

Phylum LYCOPODOPHYTA
Club mosses
Selaginella

Phylum SPHENOPHYTA
Horsetails
Equisetum

Phylum FILICINOPHYTA
Ferns
hart's tongue fern

Phylum CONIFEROPHYTA
Conifers
pine

Phylum ANGIOSPERMATOPHYTA
Flowering plants

Class HEPATICAE
Liverworts
Pellia

Class MUSCI
Mosses
Funaria

Class MONOCOTYLEDONAE
Monocots
iris

Class DICOTYLEDONAE
Dicots
horse chesnut

Appendix 3: classification of animals

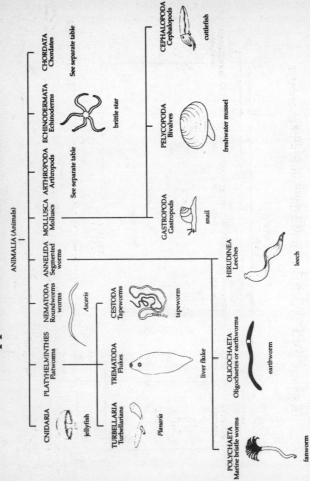

ANIMALIA (Animals)

CNIDARIA — jellyfish

PLATYHELMINTHES Flatworms
 - TURBELLARIA Turbellarians — *Planaria*
 - TREMATODA Flukes — liver fluke
 - CESTODA Tapeworms — tapeworm

NEMATODA Roundworms worms — *Ascaris*

ANNELIDA Segmented worms
 - POLYCHAETA Marine bristle worms — fanworm
 - OLIGOCHAETA Oligochaetes or earthworms — earthworm
 - HIRUDINEA Leeches — leech

MOLLUSCA Molluscs
 - GASTROPODA Gastropods — snail
 - PELYCOPODA Bivalves — freshwater mussel
 - CEPHALOPODA Cephalopods — cuttlefish

ARTHROPODA Arthropods — See separate table

ECHINODERMATA Echinoderms — brittle star

CHORDATA Chordates — See separate table

Appendix 3: classification of arthropods and chordates

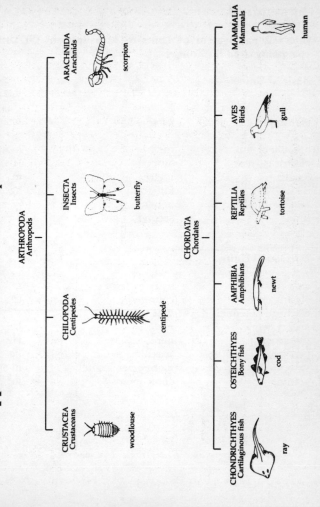

ARTHROPODA
Arthropods

CRUSTACEA
Crustaceans
woodlouse

CHILOPODA
Centipedes
centipede

INSECTA
Insects
butterfly

ARACHNIDA
Arachnids
scorpion

CHORDATA
Chordates

CHONDRICHTHYES
Cartilaginous fish
ray

OSTEICHTHYES
Bony fish
cod

AMPHIBIA
Amphibians
newt

REPTILIA
Reptiles
tortoise

AVES
Birds
gull

MAMMALIA
Mammals
human

Appendix 4

International System of Units of Measurement (SI Units) commonly used in Biology

Basic SI Units

measurement	name of unit	symbol
amount of substance	mole	mol
electric current	ampere	A
length	metre	m
mass	kilogram	kg
temperature	kelvin	K
time	second	s

Derived SI Units

area	square metre	m^2
electric charge	coulomb	C
electrical potential	volt	V
energy, work	joule	J
frequency	hertz	Hz
pressure	pascal	Pa
velocity	metres per second	$m\ s^{-1}$
volume	cubic metre	m^3

Prefixes commonly used to denote decimultiples and sub-multiples of SI Units

prefix	multiple	sign
pico-	$\times 10^{-12}$	p
nano-	$\times 10^{-9}$	n
micro	$\times 10^{-6}$	μ
milli-	$\times 10^{-3}$	m
centi-	$\times 10^{-2}$	c
deci-	$\times 10^{-1}$	da
hecto-	$\times 10^{2}$	h
kilo	$\times 10^{3}$	k
mega-	$\times 10^{6}$	M
giga-	$\times 10^{9}$	G
tera-	$\times 10^{12}$	T